P9-DWS-961

first,

we

make the beast

beautiful

ALSO BY SARAH WILSON

I Quit Sugar

And thirteen ebooks, available at iquitsugar.com

first, we make the beast beautiful

a new journey through anxiety

sarah wilson

An Imprint of WILLIAM MORROW

HarperCollins books may be purchased for educational, business, or sales promotional use. For information please e-mail the Special Markets Department at SPsalesharpercollins.com.

Originally published in Australia in 2017 by Pan Macmillan Australia Pty Ltd.

First Dey Street hardcover published 2018

FIRST U.S. EDITION

Library of Congress Cataloging-in-Publication Data has been applied for.

ISBN 978-0-06-283678-6

18 19 20 21 22 LSC 10 9 8 7

five things to know about this book

1. I'm not a medical professional. This is my personal and creative response to my condition and the research around it. But I also had three medical professionals read the book to ensure the information is responsible.
2. I've put an octopus on the cover because they are beasts that have been made more beautiful through our deeper understanding of them. Their intelligence and sentience is hard to fathom. They are driven by 500 million neurons and have a deep desire to connect and communicate with humans. The one on the cover is a Gyotaku print by Damian Oswald.
3. The scientific claims are supported as endnotes that can be found at sarahwilson.com. I acknowledge that the science in this realm is often imprecise and conflicting.
4. The format of the book is nomadic in nature. It meanders through disciplines and between polemic, didactic and memoir. Because this best reflects how I've experienced my own journey through the issue.
5. The title is derived from a Chinese proverb I came across twenty years ago in psychologist Kay Redfield Jamison's memoir *An Unquiet Mind*.

contents

THE WORM'S WAKING

This is how a human being can change:
There is a worm
addicted to eating grape leaves.
Suddenly, he wakes up,
call it grace, whatever, something
wakes him, and he is no longer a worm.
He is the entire vineyard,
and the orchard too, the fruit, the trunks,
a growing wisdom and joy
that does not need to devour.

— Rumi

the first bit

1. The first time I met His Holiness The Dalai Lama, I was invited to ask him one question. He tends to go on and on, his people told me. So one question only.

Of course I fretted. *One* question.

I was interviewing His Holiness for a magazine column I wrote in which I explored ways to have a better life. The column was one of my smarter orchestrations. Anxiety-related illness had planted me in a spot such that I was too sick to hold down a normal job, too broke to get the healing treatments I needed. So I confected a gig where I tested different ways to heal myself. Two birds, one stone.

I deliberated for days. How would I reduce things to *The* question that would provide a salve to all us Westerners seeking a more meaningful path through the fuggy, constipated, heart-sinky angst of life? The choice left my head spinning and chattering. What is it *exactly* that we need to know? Are we here to evolve into higher beings? Why are we so alone? Is there a grand scheme to our allotted eighty-five years?

When we meet a few weeks later, His Holiness kisses my hand and tosses his sandals aside. We sink into adjacent hotel room lounge chairs. I still don't have my one question. So I ask the most authentically pressing thing in that exact moment:

"How do I get my mind to shut up?"

You know, to stop the fretty chatter that makes us so nervous and unsettled and unable to grasp the "present moment" at the end of yoga classes when the instructor talks about it as though it's something you can buy off the shelf.

His Holiness giggles and blows his nose on a paper napkin, shoving it down the front of his robe like my Year 4 teacher used to. "There's no use," he tells me. "Silly! Impossible to achieve! If you can do it, great. If not, big waste of time."

"But surely *you* can do it," I say.

I mean, is the Pope a Catholic; can the Dalai Lama still his mind?

"Noooo. If I sit in a cave for a year on mountain, then *maybe* I do it. But no guarantee." He waves his hand. "Anyway, I don't have time." He has better things to do, he tells me. Like teaching altruism to massive crowds around the world.

His Holiness then tells me about his recent trip to Japan, how he hits his running machine at 3am every day and all about his anger issues (*yes, the Dalai Lama gets cranky!*). But he says nothing further about the torturous human experience of having a fretful, frenzied mind that trips along ahead of us, just beyond our grip, driving us mad and leaving us thinking we've got it all terribly wrong. It was as if the subject bored him.

I leave feeling deflated and anxious. I didn't exactly have a pearly insight for my column. But a few days later I was defending his seemingly flaccid response to my close mate Ragni and I realized what His Holiness had done.

He'd given me a response that came with a screaming, cap-lettered subtext: YOU'RE OKAY AS YOU ARE!!! He'd given me—and everyone else out there whose whirring thoughts keep them awake until 4am, trash-talking their poor souls into agitated despair—a big, fat, red-robed hug. It was perfect.

Now, a strange thing happens when you realize that some gargantuan, all-looming issue you'd been fretting over no longer needs to be fixed. You take a deep, free breath, expand a little, release your grip and get on with better things.

I suspect you might be reading these words here because you're a fretter with a mind that goes too fast, too high, too unbridled. And, like me, you might have tried everything to fix this fretting, because fretters try really, really hard at everything. They also tend to think they need fixing.

And like me you might have wondered if there's another way.

I'd like to say this up front. I write these very words because I've come to believe that you can be fretty and chattery in the head and awake at 4am and trying really hard at everything. *And* you can get on with having a great life.

Hey, the Dalai Lama told me so.

Actually, I'll go a bit further. I've come to believe that the fretting *itself* can be the very thing that plonks you on the path to a great life.

2. When God was handing out *The Guidebook to Life* I was on the toilet. Or hanging out nappies for Mom. I was, I believe, the only person on the planet who missed out.

The first time this realization came crashing down on me I was fifteen, crouching in an Asian-style squat behind a curtain in a Canberra shopping mall waiting to see if I'd won the inaugural Face of Miss Gee Bees modeling competition.

Miss Gee Bees was the teen section of the now defunct Grace Bros. department store behemoth, should you be too young to know.

A few months earlier a matronly fitting room attendant had stopped me as I flicked through her bra rack and asked if she could take a photo with her point-and-click. "Yeah. Okay," I said and half-smiled, half-frowned for the camera. I got a letter two weeks later inviting me to attend the finals being held at the mall's center stage. Up for grabs was a modeling contract, a *Dolly* magazine shoot and a bra and underwear package.

The other finalists chat and laugh as we wait for the judges' announcement. They're glossy and cheerleader-y and all seem to be wearing the same Best & Less stilettos and black lycra micro-dresses that they keep adjusting over their bottoms, but without bending over. Because to do so would muck up their hair-sprayed quiff-fringes, a few strands combed forward over their eyes.

I hadn't got the Robert Palmer memo.

I'm wearing an ankle-length white poly-cotton peasant dress with beige slouch socks and worn tan Sportsgirl brogues with splits in the soles. It's a bit Linda Kozlowski in *Crocodile Dundee*. A little bit *Out of Africa*. I'd borrowed the dress from a girl at school and I'd carefully hand-stitched the princess-line seams in a few centimeters to fit. Dad superglued the splits in my shoes and dried them on the hot water tank overnight.

I'm feeling nervous. And, oh boy, so terribly alone.

Also, this had just happened. On my second run on the catwalk for the Saturday morning shopping crowd I'd spun in front of the judges. All eyeballs were on me.

And. Then. Time seemed to stop and the world went silent like it does just before a bushfire.

And. Then. A tightly packed wad of toilet paper dropped from under my dress and landed with a light *pfft*, right in front of the judges.

As an awkwardly undeveloped teen, I'd do this thing where I'd stuff tissue or toilet paper into the sides of my underpants to give me hips where I had none. I wore jeans mostly, and would stick the toilet paper in the pockets, wash them and then bake them in the sun, creating papier-mâché insta-curves. I'd also wear two—sometimes three—T-shirts at a time, rolling the sleeves up over each other, and football socks with ankle boots to bulk out my undersized frame. I was an optical illusion of womanly shape that had to be carefully, anxiously, constructed each morning.

From the back of the crowd Dad whooped, "You little *beauuuuty*!" like he did at sports carnivals when my brothers and sister and I ran into the home stretch, no matter our placing in the pack.

I scampered off behind the curtain.

Was I mortified? Ashamed? No. This wasn't the issue. The ordeal had instead triggered a panic, an overwhelming and lonely panic of the most fundamental kind.

I was breathless and alert in the car on the way home with my second-place bra and underwear package. Entirely unanchored, dangerously adrift. In this moment I fully believed that I didn't get it. I didn't "get" life. And everyone else on the planet did. They'd got *The Guidebook*. They got the missive that showed them how to interject in a jokey conversation. They got the instructions for choosing the right career path. They seemed to somehow know why we existed. Shit! Shit! How was I going to get through this thing called life?

But! One of the dear, dear things about getting older, is that it does eventually dawn on you that there is no guidebook. One day it suddenly emerges: *No one bloody gets it! None of us knows what we're doing.*

Thing is, we all put a lot of effort into looking like we did get the guide, that of course we know how to do this caper called life. We put on a smile rather than tell friends we are desperately lonely. And we make loud, verbose claims at dinner parties to make everyone certain of our certainty. We're funny like that.

3. Stephen Fry wrote in *The Fry Chronicles* that behind "the mask of security, ease, confidence and assurance I wear (so easily that its features often lift into a smirk that looks like complacency and smugness) [is] the real condition of anxiety, self-doubt, self-disgust and fear in which much of my life then and now is lived."

Two things about this.

THING 1. IT'S THE MOST INCREDIBLE RELIEF TO KNOW THAT WE'RE ALL WEARING masks . . . and to see them slip on others. Oh, sweet Jesus, we're not alone! We're in this together! It's not a mean-spirited *schadenfreude*; it's the ultimate connection. Really it is. My beaut and brutally frank mate Rick rang and asked me one morning, "Darl, why exactly are you writing this book?"

"Because I can't help it and because I'm sick of being lonely," I replied. Then I quoted something I'd read that morning from philosopher Alain de Botton's *The Book of Life*: "We must suffer alone. But we can at least hold out our arms to our similarly tortured, fractured, and above all else, anxious neighbors, as if to say, in the kindest way possible: 'I know . . .'"

"Good," Rick said and hung up.

Thing 2. When you realize there's no guidebook, an opportunity suddenly presents itself. If no one knows what they're doing, if there's no "right" way to do life, then we can surely choose our own way. Yes?

4. My beautiful brother Ben recently asked me over the phone, "Remember that time you got stuck on the bus because of that woman's perfume?"

Nope. But if Ben, the family elephant, said it happened, it did. Ben's sixteen months younger than me and I realize just now that he's been my ballast over the years with his gruff, "Sarah, just don't worry about it" sturdiness. The Mindy to my Mork.

Apparently I was so distressed by the stench from the lady sitting next to me I'd covered my face and missed several bus stops. Perfume has always made me anxious. I was six.

I've been anxious for a long, long time. I don't know when or how it kicked in, but I don't remember a time without it.

5. I was diagnosed with childhood anxiety and insomnia at twelve, then bulimia in my late teens, then obsessive-compulsive disorder (OCD) shortly thereafter, then depression and hypomania and then, in my early twenties, manic depression, or bipolar disorder as it's now called.

I've seen about three dozen psychiatrists and psychotherapists and spiritual healers, generally twice a week for years at a time. I was medicated from seventeen until I was twenty-eight with anti-epileptic, anti-anxiety and anti-psychotic drugs. I've waded through cognitive behavioral therapy (CBT), neuro-linguistic programming (NLP), hypnotherapy, Freudian analysis, spiritual coaching and sand play. For long, lonely slabs I've had to step out of the slipstream of life, missing school, dropping out of university twice, quitting jobs and unable to leave the house for up to a year at a time. Also twice.

I can now tell you it was all anxiety. All of it. Just different flavors.

But at twenty-seven I decided to go my own way. I was living in Melbourne, writing restaurant reviews and celebrity features for the Sunday paper. I also wrote a weekly opinion column. I'd write it Thursday night and had the most marvelous time, under the pump, with an outlet for my thoughts on homeless people, feminism and the reasons why men always power-walk in pairs. I'd recently split from my first boyfriend and was living with a fun artist in a South Yarra terrace that was to be demolished in coming months. We wrote on the walls, ivy grew through the kitchen, we cooked stew. And I was on a conscious mission to explore sex. I came to sex late and had only had one sexual partner. I was ready to play; it was a fun experiment and one not based on pain or compromise. Things felt aligned and touched by some rippin' flow.

And so I broke up with the psychiatrist who was my last for a very long time. I presented her with a dot-pointed rationale of why I had to go my own way. "I am ready," I told her. "This is the real thing, now. Life ain't no run-up, a dress rehearsal," I said. "I'm ready for the work. It's just hard work, right? I can do hard work. It's a matter of firing the fuck up." She shook my hand as I left her dimly lit office overlooking Melbourne's Albert Park. I appreciate, now, that I was probably riding a slightly manic upswing.

Six months later I had used up the last of my medications. They'd run out, one by one. And I'd simply chosen not to repeat the prescriptions.

Despite appearances, this was not a monumental fork-in-the-road-never-turn-back moment. That's the thing with my important life moments, they always seem to emerge slowly, like a Polaroid picture. I suspect few people have instant-capture aha moments. Especially those of us ensconced in the nebulous realm of anxiety where discernible lines between normal and neurotic cease, at some point, to exist.

That said, I think my adult journey, the one I'm sharing in this book you're holding, began as I left my psychiatrist's office on that late autumn morning. I remember the soft light. I remember doing a fist-pump as I walked to the tram stop. I was making up my own rules for managing what everyone insisted on calling an illness and I knew I was ready to live them out. I get asked how I did this. I can only say that I chose. I made the decision and then I committed, motivated predominantly by the fact that, frankly, nothing else had worked. I've spoken to a lot of functioning neurotics over the years and they tell me the same. You choose. You might not even know why, but you do. You commit. Then you do the work.

Oh, yeah. Then you falter. And fuck up. And go back to the beginning.

In my mid-thirties my mania flared again. And my obsessive-compulsive disorder. I've wrestled with OCD since I was eleven or twelve. I have to tap things and check things, and wash my hands, to a count of three. It's a night-time ritual only. I tap light switches and doors and bathroom taps after everyone has gone to bed and I check—to a count of three, in multiples of three—for things under my bed. As a kid I counted pretty much everything in threes—cracks, drips, turnings of my pillow to the cool side when I couldn't sleep. I know when I'm getting worse. My counting goes from sets of threes to sets of fours and fives.

At thirty-five I was also suicidal for the second time in my life. I was unable to leave the house or go to work for nine months. Everything unraveled again.

– **The first time was when I was twenty-two, but I'll cover this a bit later on.**

I've since gone back to therapists. I've gone back to medication. And then gone off it again. I have anxiety attacks in batches throughout the year. I keep Valium in my bathroom. Just in case.

But this journey is what I do now. I bump along, in fits and starts, on a perpetual path to finding better ways for me and my mate, Anxiety, to get around.

It's everything I do.

As someone wrote to me on my blog a few years ago, "Sarah, you're all striving, no arriving." Yes, and I think this is the point. I've written more than 1500 posts on my personal blog, and hundreds (thousands?) of agonized-over columns for eleven different magazines and newspapers over a twenty-two-year writing career in which I try out different ideas and life hacks all geared in one way or another at . . . what? . . . understanding my anxiety. It's been a rather self-serving career trajectory.

Where do I land? Modern medicine has certainly preferred dealing with my various conditions as individual diagnoses—the bipolar, the anxiety attacks, the obsessive-compulsive disorder, the insomnia, as well as the host of autoimmune diseases I've developed along the way (and never shall thy psychiatrist and endocrinologist meet!). But me? I think all the diagnoses boil down to anxiety. That is, an itchy sense that things are not right, a buzzing dis-ease. Whatever doctors want to call it, the feeling is the same: it's that gut-twisting, grip-from-behind, heart-sinky feeling that winds me in tighter spirals and makes everything go faster and with so much urgency and soon enough I'm running down a steep hill faster than my poor spinning legs can carry me.

I'm aware many doctors may disagree with this notion, and I could indeed be wrong. But I reckon it's time we explored the idea. It's time we had the conversation because I think many of us are feeling the itchiness of something missing from the issue.

My qualifications for writing this book, then, if this matters to you, is that I'm a committed striver. I'm strapped in. Doing the work and keen to start the conversation.

I should also point out that I don't have an answer to . . . any of it. You should probably know this eleven pages in. This book doesn't take a linear path to salvation. Nope, it meanders through a series of explorations. I take off my mask and share my not-knowing.

But, dear reader, I ask you, do you feel, in your heart of hearts, that fixing your anxiety is the answer? I ask this of anyone with the kind of low-to-medium anxious buzz we're all feeling, as well as those of you with a diagnosed anxious condition. Because the question is equally relevant. Do you think it might be lovelier if we bundle up our uncertainty, fear, late-night overthinking and kooky coping habits, tuck them gently under our arm, and see where they take us?

This might not sound like the most "grab a highlighter and mark out the wisdoms" premise for a book. But let's see how it goes.

6. *The Journey* by Mary Oliver goes straight to it. I don't generally like long quotes and poems in books. They clutter the flow. But please, do yourself a favor and read it in full (it's just on the next page), and absorb it, so we can all start on the right note.

THE JOURNEY

One day you finally knew
what you had to do, and began,
though the voices around you
kept shouting
their bad advice—
though the whole house
began to tremble
and you felt the old tug
at your ankles.
"Mend my life!"
each voice cried.
But you didn't stop.
You knew what you had to do,
though the wind pried
with its stiff fingers
at the very foundations,
though their melancholy
was terrible.
It was already late
enough, and a wild night,
and the road full of fallen
branches and stones.
But little by little,
as you left their voices behind,
the stars began to burn
through the sheets of clouds,
and there was a new voice,
which you slowly
recognized as your own,
that kept you company
as you strode deeper and deeper
into the world,
determined to do
the only thing you could do—
determined to save
the only life you could save.

— Mary Oliver

Don't you reckon?

because no one knows . . .

7. So here's the thing. A lot of us are anxious. Many of us haven't been diagnosed as such, or even worked out if our particular flavor of anxiety constitutes a problem. But we know we're anxious. More anxious than we should be. When I mention I'm writing a book about anxiety, everyone (and I mean every single person) suddenly goes a little wide-eyed. Drops their tone a little. Leans in. And tells me . . . "Everyone these days seems to have it, hey."

We're told that globally one in thirteen people suffer an anxiety-related illness. Some studies tell us that one in six of us in the West will be afflicted with an anxiety disorder at some stage in our lives, making it the most common officially classified mental illness. For men, anxiety is even more common than depression—one in five men will experience anxiety at some point. But of course these are only the stats for those whose anxiety crosses the line to become a diagnosed disorder. There's no accurate gauge for the number of people who are feeling the frenetic toll of modern life but somehow manage to keep on keeping on without presenting to a doctor, though Google search rates can sometimes give a real-time picture of such things. And, sure enough, searches for anxiety are up 150 percent in the past eight years. Searches for "anxiety at night" have risen nine-fold.

A growing number of conditions come under the umbrella term of anxiety, in the medical sense of the word. The latest (fifth) edition of the Diagnostic and Statistical Manual of Mental Disorders (DSM), the diagnostic handbook used by mental health professionals in most Western countries, lists thirty-seven different disorders under anxiety, including social phobia, specific phobias, panic disorder, generalized anxiety disorder (pervasive and chronic worry about a variety of everyday issues), separation anxiety, obsessive-compulsive disorder (OCD) and post-traumatic stress disorder (PTSD). (I should note that since 2013 the American Psychiatry Association has assigned OCD its own separate category.)

Anxiety was first classified as a mental disorder in 1980 in the third edition of the DSM.

Yep, you read right: 1980.

And before then? Were none of us anxious?

For nearly three centuries—pretty much since Descartes separated our heads from our flesh and bones—the idea that our emotions could impact our physical health remained scientific taboo. In 1950 only two books had been written on the topic. Two. In the whole entire world. (Freud wrote about it eighty years ago in *The Problem of Anxiety*, and Søren Kierkegaard ninety years before him in *The Concept of Anxiety*.)

We officially became "stressed" from 1950, when the term was invented by an Austrian–Canadian physician who worked tirelessly to popularize the term with a one-person campaign to get the word into dictionaries around the world. An etiology I just love.

Several dozen more books and academic papers then appeared on the subject. In the main, though, the mind-body split persisted and anxiety was regarded as an everyday condition (we all get anxious, right?) that some of us are just too weak to handle. Women who got anxious were slapped with the hysteria diagnosis. Men self-medicated with drugs or alcohol or went into man caves. And we all got on with things.

Today there are countless theories as to the genesis of anxiety. All of them are worth understanding. Childhood trauma is one. In Sue Gerhardt's seminal book *Why Love Matters: How Affection Shapes a Baby's Brain*, it's argued that when you grow up in conditions of emotional vulnerability (anything from extreme abuse to having a parent who was busy with other children—hello, the Anxiety of Being An Eldest Child!) it can shape how you cope emotionally for the rest of your life. British TV clinical psychologist Oliver James (author of the brilliantly titled parenting book *They F*** You Up*) argues that childhood trauma sees you grow up in a state of permanent "red alert" that deregulates your brain chemistry. It's firmly nurture, not nature, in his book.

Low levels of certain chemicals in the brain (namely serotonin and norepinephrine) have also been implicated. Our brains process information about what's happening inside and outside the body via the nervous system—a network of 100 billion neurons that relay information via electrochemical signals. Serotonin relays messages related to mood, sexual desire and function, appetite, sleep, memory, learning and social behavior. The crude theory is that when we don't have enough serotonin, the information doesn't get through.

More recently, it's been found that another neurotransmitter, anandamide (a name that comes from the Sanskrit word *ananda* for joy or bliss), helps our brain communicate happiness, ease and comfort. Those of us with low levels of this message emitter tend to get nervous. Or so goes the theory.

Even during the writing of this book, I've been exposed to fresher explainers.

Recent research has shown that anxiety is more common in people with autoimmune (AI) diseases, illnesses that occur when the immune system gets deranged and attacks different parts of the body. I have Hashimoto's, an autoimmune disease of the thyroid, a gland in the neck that controls everything that makes you a conscious, sentient being: metabolism, breathing, heart rate, the nervous system, menstrual cycles, body temperature, cholesterol, blood sugar, mood, sleep . . .

No one can explain the connection between AI and anxiety, in part because the genesis of many autoimmune diseases is not known. Some suggest that the uncertainty of living with an autoimmune disease is a possible factor—simply not knowing what the hell is going on with your body would drive anyone to agitated despair.

A more promising, and compelling (to me), explanation is that AI, like other inflammatory diseases, lead to high levels of circulating cytokines, which have been shown to affect how our neurotransmitters communicate. Indeed, increasing evidence links anxiety to a bunch of inflammatory diseases such as atherosclerosis, metabolic syndrome and coronary heart disease. Clinical trials have shown that adding anti-inflammatory medicines to mood medications not only improves symptoms, but also increases response rate. The fact that "normal," healthy people can become temporarily anxious after an inflammatory vaccine—like typhoid—lends further credence to the hypothesis.

Possibly the most exciting work is being conducted on the link between gut health, inflammation and anxiety. As you've no doubt heard, we have a whole community of bacteria in our digestive tract—our microbiome—which not only plays an important role in our metabolic and immune systems, but also our nervous system. Recent research suggests that these microbes may influence emotional behavior, pain perception and how we respond to stress.

When the balance of "good" and "bad" bacteria is out of whack (dysbiosis, caused by such things as poor diet, medications, antibiotics, allergens, parasites, fungal overgrowth), it can trigger a cascade of inflammatory molecular reactions that feed back to the central nervous system, causing inflammation in the brain. And it's this inflammation that messes with our neurotransmitters, leading to anxiety. Put simply, if you have fire in the gut, you have fire in the brain.

Researchers have found that folk who eat more fermented foods (which contain gut healing probiotics) have fewer symptoms of social anxiety. Another study found that eating a mere yogurt (I say "mere" because the commercial stuff contains only small amounts of the beneficial bacteria touted on the front of the tub) twice a day for a few weeks changed the makeup of the subjects' gut microbes, and this led to the production of compounds that modified brain chemistry.

In the final weeks of writing the first draft of this book, I found out that I have a number of genetic issues that likely played a part in my anxious history. I have a defective MTHFR gene, which can affect serotonin availability and uptake. It's a new area in the debate, so I'm still learning about its implications. My hormone clinician, Leah, suspects that I was also born with an inability to produce enough glutathione, a key antioxidant that helps eliminate heavy metals from the body. As a result, I have mercury levels that are literally off the scale (as in, beyond the graph that the labs provided). High levels of mercury, of course, are directly linked to anxiety. "Mad hatters" went nutty from the mercury used in felting work. Leah reckons these genetic variables may have been "switched on" by epigenetic factors such as stress, illness and environmental pollutants.

Nascent research published in *Nature Neuroscience* also hypothesizes that phobias (an anxious manifestation, as per the DSM) may be genetic "memories" passed down from our ancestors and mediated by epigenetics, which may help to explain why people suffer from seemingly irrational phobias that probably made sense at some point in our evolution (like fear of germs).

But none of the above factors consigns a person to an anxious diagnosis. You can have all of the above and never have an anxious day in your life.

8. You know what else happened in 1980, just prior to anxiety being formally recognized and diagnosed? The first anti-anxiety drugs were manufactured. Which begs, was anxiety "invented" in 1980 to fit the drug? Just a question, just a question, people . . .

Oh, and there's this.

In 1994, as Scott Stossel notes in *My Age of Anxiety*, the term "social anxiety disorder" had only appeared fifty times in the media. By 2000 it was part of our lexicon and studies were indicating that 10–20 million Americans were afflicted.

What might have triggered such a dramatic switch? Were we all suddenly being forced to go to more crowded parties in the late '90s? Or to attend more group workshops involving those awkward trust exercises?

Or could it have been the widespread awareness campaign

launched by the Social Anxiety Disorder Coalition (SADC) in 1999, just one year earlier? This particular campaign featured glum-looking folk accompanied by the truly awful tagline, "Imagine you were allergic to people" and got blanket coverage across America, on the backs of buses and on freeway overpasses.

But sit tight! There's more.

The SADC turns out to have been a partnership with SmithKline Beecham (which later became drug giant GlaxoSmithKline), which—hello!—had just released the world's first pill, Paxil, approved to treat—you guessed it!—social anxiety disorder. Obsessive-compulsive disorder and bipolar disorder have similar drug-first histories.

Now, let me say very clearly. I'm not suggesting anxiety is a confection. I live it out viscerally, daily. So do the estimated 14 percent of Australians whose lives are disrupted by OCD, social anxiety, post-traumatic stress and other anxiety-related illnesses.

Like David Beckham, who gets sent over the edge if things in his life aren't in a straight line, or in pairs: "I'll put my Pepsi cans in the fridge and if there's one too many then I'll put it in another cupboard somewhere . . . Everything has to be perfect." And Whoopi Goldberg, who avoids flying. I'm not sure why I've inserted these celebrity anecdotes. Perhaps just to lend color and weight to the fact that anxiety is the sixth leading cause of disability in the world.

So yes, this stuff is real. And learning about it and carefully considering the various diagnoses and suggested causes and treatments is important. Indeed, it's vital to the journey. But, I guess my brushwork here is broader. I'm plonking down the above factlets because they pertain to many of the quandaries every anxious person I know has had to face.

Am I really mentally ill? Disordered? Defective?

Or am I just weak of character and just not trying hard enough?

Does taking medication alter who I am? Am I less authentic for it? Is it "unnatural"?

And am I clinging to the "chemical imbalance" theory because it absolves me of blame and the science-y-ness promises a neat fix?

Or are my neuroses fair enough given the state of the world today? Is my fear of crowds, confined spaces, financial ruin, being touched, etc., a reasonable evolutionary response, albeit one that has got a little bit off kilter?

I've asked these questions for years. We have to, we anxious folk. The existing medical theories fail to answer them.

You ask the same? Well, you're probably just as interested as me to learn that despite grand efforts to classify anxiety as a distinct disorder, there is no diagnostic process that actually works. No technique has been established that can determine the line in the sand where normal stress and fear become neurotic anxiety, or at what point your whirring thoughts can be explained as a chemical flaw in the brain instead of a character flaw (the former, of course, being far more "forgivable" and, of course, "fixable").

In fact, studies of the DSM's diagnostic guidelines have found that when two different psychiatrists used the same edition of the DSM to diagnose the *same* patient, they get a consistent result only 32–42 percent of the time.

Again, terribly vague. In Australia, just to confuse matters further, doctors work to both the DSM guidelines and a set of classifications set out by the World Health Organization (WHO), which are, not so conveniently, at odds with each other. To make matters worse, public hospitals work to an outmoded version of the WHO classifications when coding cases. Lost? Yeah, so are most medicos.

I think many of us when we're young find the chemical imbalance rationale comforting. Which is fair enough. For me it provided a shelf on which I could place things for a bit until I could process them. To be told that we have an illness that is not our "fault" relieves some of the doubt and uncertainty, and absolves us of the guilt we feel that we should be able to cope better. Which in itself turns down the anxiety dial.

Back in my late teens, doctors told me I had a serotonin deficiency in my noggin.

I'd moved out of home not long after my little brother was born and there was chaos all around me in the group house I shared with three older students. One of them loved to vacuum and practice guitar late at night; another self-harmed in her annexed bedroom off the kitchen (she claimed it was the yellow paint on the ceilings that caused it). I was studying law and politics. And working three jobs.

I was not sleeping and I was tapping and I was counting and I went on wild, wailing runs in the bush reservoir nearby at 4am when I couldn't sleep.

And I'd sing Violent Femmes' "Blister in the Sun" (*da doonk, da doonk*) over and over.

I let out a big sigh when the psychiatrist explained my brain in easy-to-illustrate electrical circuitry terms. Finally, I had a tangible and apparently touch-it-with-a-scalpel reason for this mess in my head! More importantly those around me had a reason. It was hereditary, unavoidable and my parents weren't to blame (at least not beyond the transferral of their genes).

Certainly it all seemed to be an unavoidable instance of genetic misfortune. My grandmother on Dad's side was institutionalized for manic depression. She told me on a four-hour bus trip from Canberra to Sydney when I was about fourteen that she'd had electric shock treatment four times against her will. Dad was very young at the time and the details were always kept quiet. Mom's mom suffered from terrible anxiety and was hospitalized when Mom was young. Again, these things weren't discussed much and the details are sketchy.

This whole corner of the debate is likely to get a rise from some of you. Rapper and sensationalist YouTube dude Prince Ea waded into this territory recently and, boy, did he cop it in the comments section of his video from anxious thousands around the world tired of being blamed for their condition. Prince Ea said (to a motivational beat) that we are not our anxiety or depression. No, we are the sky, and anxiety and depression are but clouds that pass through us.

I see what he was getting at. I am not my sickness; I have a condition that can wander all lonely and cloudlike into view from time to time. I (the whole me) can choose to sit back and witness the clouds, let them be, let them pass. *Pfft.*

However, much online outrage ensued with many arguing their illness was most definitely medical and not a passing thing. "Would you tell someone who is diabetic, 'You are not diabetic, you're just choosing to focus on insulin and blood sugar levels'?"

Yep, once again, I see what they were getting at. For some of us, it does get to the point where the bloody clouds take over the sky. There is nothing left but black clouds. It *becomes* medical.

That said, since I was first presented with the "chemical imbalance" explanation, it's been found to be largely unproven and increasingly regarded as incorrect given how little we actually understand about how the brain works. I was pointed to the U.K. Royal College of Psychiatrists' website recently. On the Treatments and Well-Being page it tells us (bless them for being so frank), "We *don't know for certain,* but *we think* that antidepressants work by increasing the activity *of certain chemicals* in our brains . . ." The italics are mine. You have to roll your eyes a little, yes?

Besides, the quandary remains: is a chemical imbalance the cause, or the effect? Did my grandmother's anxiety instead *cause* a drop in her serotonin levels?

I admit I feel nervous about questioning the medical model when I'm not a doctor (although history shows that it's often laypeople who draw attention to gaps in science). I consulted a number of psychiatrists and heads of certified mental health organizations to ensure that my questioning was neither misleading nor harmful. I was told repeatedly over animated coffee chats around the country that it was not. Dr. Mark Cross is a psychiatrist and board director of SANE Australia. He claims openly that modern psychiatry at times suffers from overmedicalization and often a lack of informed consent when patients are given drugs without being given the full picture, including the fact that drugs are not curative in many instances. "We have not cracked the cure, yet patients are often given drugs without being told this." He goes as far as to say that the medical model can sometimes cause harm, which goes against the Hippocratic oath (which he took during his medical training).

And so you take all these unanswered questions and deficiencies to your shrink or to the self-help aisle or the internet.

One of the psychiatrists I saw in my early twenties, a grandfatherly man with a snow-white beard, conceded that yes, we're talking about an inexact science here. He added this warning. "Be careful, Sarah. You're very seductive." He was effectively telling me that less wary doctors than he might be driven to abandon their treatment plan for me in the face of my robust questioning of the science. Which only fueled my cynicism.

Ready to roll your eyes again? In 2011, the dude in charge of the DSM was asked by a journalist if the line between normal human response to threat and certifiable "disorder" might not be as distinct as his didactic tome makes out. His response was—you can probably guess it—"I don't know." In another interview he confesses that there are no biological markers for anxiety disorders.

And you might scream back, as I did when I reached this point, "But we're not just talking about a bunch of cells and synapses on a screen here; this is my goddamn sense of self, the stuff that makes me, me!! Are we really doing this experiment, people?!"

So my point is this. Take on board all the theories. But given no definitive causes, diagnoses or treatments have been found yet, why not see this as an opportunity? An opportunity to define anxiety as something other than a problem or disorder that has to be fixed as such. U.K. *Guardian* columnist Oliver Burkeman, who dedicates his column inches to questioning self-help culture, asks whether our focusing so heavily on defining the problem tacitly creates the problem, namely that we're broken and require fixing. "Perhaps the problem, sometimes, is the notion that there's a problem."

I must emphasize: learn, learn, learn. And be open to it all. This is pretty much the raison d'être, the joy of this journey.

Dr. Mark Cross agrees: "Just because you're diagnosed with anxiety, doesn't mean you have a problem, one that needs medication."

So, I ask, could we go our own way? Could we play a little with not having a problem? Could we be so bold?

9. Although, let me emphasize, pills and shrinks and behavioral therapies each have their place. For sure. So does a medical diagnosis. As I say, even just addressing the issue with an external party is invaluable. It can get you started on the journey. And, again, for those new to their anxiety—especially teens—a diagnosis can be a safe place to plant things until you have the wisdom and learning to take you into deeper understanding.

Nicely, the word *diagnosis* itself comes from the Greek *diagnosi*, which means "to know through."

The year before, as I launched into writing this book, my insomnia bottomed out after almost a year of trying to get by while berating myself relentlessly for not coping better. Bugger the root cause of my anxiety! Bugger the vested interests of the drug giants! Bugger whether the chemicals in my brain were awry or not. I. Was. Not. Coping.

The lack of sleep night after night was causing my autoimmune disease to flare up terribly and I was struggling to function most days. I couldn't hold a thought for longer than a few seconds, I was totally irascible and every nerve ending hurt. I was in pain—an inflamed, aching pain that I describe to those who care to know as someone running their nails down a blackboard and rubbing your nerve endings with a scrubbing brush, simultaneously. I was existing like this five or six days a week. My former business manager kept having to take me aside and explain the impact my frazzled state was having on the team in the office. Every human irritated me. Unlike in the past, when I worked on my own, I couldn't squirrel myself away from the world, saving them from my subhuman presence. I was now highly accountable.

It was when a friend confided that she couldn't run her fashion styling business without seeing a therapist every week that I was reminded of a sensible commitment I'd made back when I first went my own way at twenty-seven. I'd promised myself I'd get help if I needed it again. I went back for anti-anxiety tablets, which I took for ten months. And then I went back for psychoanalytic psychotherapy, an old-school style of therapy that focuses on understanding and expressing feelings, which I'd abandoned for many years in favor of the more modern cognitive treatments, which teach techniques that control said feelings.

I saw Dr. H for six months before he dismissed me, finding me to have stabilized enough to venture back out on my own.

Like I say, it's a bumpy journey.

10. Even today, truth be told, I'm not entirely sure I have an anxiety disorder. As such. Or if I'm just terribly deficient at coping with everyday living. You too? Which leads me to point out the first of many cruel ironies in the anxious struggle. There are a few.

The curious nature of anxiety is such that it defies its own diagnosis and treatment. — **cruel irony #1**

Let me explain.

Anxious behavior is rewarded in our culture. Being high-strung, wound up, frenetic and *soooo busy* has cachet. I ask someone, "How are you?" and even if they're kicking back in a caravan park in the outback with a beer watching the sunset, their default response is, "Gosh, so busy, out of control, crazy times." And they wear it as a badge of honor.

This means that many of us deny we have a problem and keep going and going. Indeed, the more anxious we are, the more we have to convince ourselves we don't have a problem. This is ironic, or paradoxical. And it seems awfully cruel. I read an interview with a clinical professor of psychology and psychiatry at Weill Cornell Medicine who said it takes 9–12 years for women to get a diagnosis for anxiety. I presume it would be just as long, possibly longer, for men. We suck it up when we feel anxious and soldier on until, well, we tip over the edge and our anxiety turns pathological and medical. Flipside, depressed behavior—slovenliness, unproductiveness and suck-holey gloom—is something we abhor. Thusly, depression is an issue. And, thusly, we have lots of structures in place to identify it and treat it.

Depression is stigmatized, anxiety is sanctified as propping up modern life, which ironically sees depression treated as a legitimate illness, and the anxious left in a cesspool of self-doubt and self-flagellation for not being better at coping with life. And so we buy each other Keep Calm and Carry On mugs as though that's something you can just do.

But it gets worse, you see. We then try to cope by revving up the angst, don't we? We use coffee and fast-speak and sugar and staying back at work longer. We grind harder. Try harder. Think harder. We *should* be able to *work* our way through this. We think this is what will fire us up out of our funk and get us back on our game. It's a self-perpetuating pain—we use anxiety to fight our anxiety.

But we often don't see it until it's too late. I suspect many of you with anxiety who are reading this struggle to see it clearly. And struggle to treat it appropriately, with candlelit baths and "me time" and cortisol-calming medicinal foods. I do (struggle) and I often don't (do candlelit baths when I'm nuttily fast in the noggin).

For years I saw my life as a stacked spiral of dominoes. Until I realized a Jenga stack was probably a better metaphor. I was wholly convinced that if I removed my bull-at-a-gate approach, even just a few struts, the whole structure would topple. I was told to "back off." To "just relax." I dismissed such notions because they only induced further anxiety. I felt I'd be nothing without my anxious drive and when I felt it sag a bit I'd panic. I'd sturdy things again with a stern talking-to. "Fire up, Sarah." I'd rev up my adrenals with punishing runs and double-strength coffee. I could not let this whole game fall over.

I was rewarded for doing so with better jobs and lots of admiring looks. Until a gentle zephyr happened by and knocked the whole rickety structure to the ground. (We'll get to this in a bit.)

We also need to recognize—which many doctors don't—our anxious behaviors are so often the *solution* to our problem, not our problem. Take bulimia. I struggled with it in my teens and twenties. I was lucky to have a mom who'd never dieted in her life and loved food. Food was not stigmatized in our house. For me (and I suspect for many others) bulimia wasn't about food or body image directly (although the fact I expressed my anxiety via my body means I can't fully escape the connection). It was about controlling my anxiety. I jammed raw oats and Vita-Brits (anything heavy and gluggy) down on top of the fluttering I felt in my guts. It dampened it for a bit, suffocated it. Guilt, despair, disgust would build, not so much from the gluttony and fear of weight-gain (although, yes, it was connected), but from the anxiety working its way back in again. So I'd purge. I'd purge the anxiety. The bulimia was my solution, in the absence of anything better at the time, and I never, ever spoke about it because I didn't want anyone to get it into their heads to take it away from me.

And don't try telling me it was the modeling work I did when I was younger that caused it. It was only while I was modeling that the bulimia backed away, for reasons that had a lot to do with the security that came from my handsome paychecks, and the fact that it took me away from what was then a turbulent home life.

A BRIEF NOTE ON HIGH-FUNCTIONING ANXIETY

Many of us with anxiety don't look like we've got a problem because outwardly we function ludicrously well. Or so the merry story goes. Our anxiety sees us make industrious lists and plans, run purposefully from one thing to the next, and move fast up stairs and across traffic intersections. We are a picture of efficiency and energy, always on the move, always *doing*.

We're Rabbit from *Winnie the Pooh,* always flitting about convinced everyone depends on us to make things happen and to be there when they do. And to generally attend to *happenings*.

But beneath the veneer we're being pushed by fear and doubt and a voice that tells us we're a bad husband, an insufficient sister, we're wasting time, we're not producing enough, that we turn everything into a clusterfuck. Sure, we look busy, but mostly we're busy avoiding things. So we tie ourselves up in stupid paper-shuffling-like tasks that shield us from ever getting around to the important stuff. Or the tough stuff.

P.S. *Clusterfuck* is a military term from the 1960s. It refers to a chaotic, complex situation where everything seems to go wrong.

And, yep, we're the ones who send out random texts suggesting we all catch up for dinner next week. We're also the ones who cancel at the last minute. And who simply do not pick up the phone for days (weeks?) when it gets too much. We go underground. We remain single for decades. And everyone just assumes we're too busy and high-functioning for such things.

On a blog post entitled "When Social Anxiety Looks Like Talking Too Much" I read about one girl's battle with the seemingly inconsistent appearance of her anxiety. "To everyone, I'm just the girl who talks too much and tries too hard. When really, I'm just trying to quiet this battle in my mind for the hour."

I'd add that, in such instances, we'd love everyone (someone?) to see that we absolutely do not have our shit together. And to come and tell us they've got this one. Even for five minutes.

The more anxious we are, the more we'd really love someone to come and take the load off us and help us cope for a bit. This presents us with another cruel anxious irony, doesn't it:

The more anxious we are, the more high-functioning we will make ourselves appear, which just encourages the world to lean on us more. — **cruel irony #2**

Anxiety . . . it's befuddling and clusterfucky for everyone involved.

11. Sometimes, particularly when I give myself a hard time about taking medication to sleep (variously, valerian, melatonin, paracetamol, cold and flu tablets, Valium and Seroquel), I recite to myself the chorus of John Lennon's "Whatever Gets You Thru The Night" (*s'alright, s'alright*). I find the certain *c'est la vie*-ness of it brings a lightness, an expansiveness, to things when I get into one of my fretty funks. And we need to find lightness where we can (I think it's why so many comedians make jokes about their neuroses, and even why so many neurotics become comedians). I tend to repeat Lennon's very forgiving line over and over. This is because I'm really rather, literally, obsessive-compulsive.

12. Yeah. So. Let's talk some more about insomnia. I've had it, on and off (mostly on) since I was about seven. As a kid, I was superaware that Dad couldn't sleep and I'd hear him in the middle of the night from down the hall, agitated, freaking out to Mom about how he'd cope in the morning. I took this on, as an eldest daughter just does. I realize I also blamed myself for Dad's insomnia (something—helpful or otherwise—you learn after a few decades of therapy).

Insomnia is anxiety's spiteful bedfellow. Anxious people desperately need sleep, yet their condition ensures they are denied it.

cruel irony #3 — The less you sleep, the more anxious you get, the less you sleep . . . and so on.

Indeed, ironic. And in a not-so-fair kind of way. Yet this is all medical fact. Neuroscientists at University of California Berkeley have found that sleep deprivation fires up the same abnormal neural activity seen in anxiety disorders. Worse, the already-anxious are more affected by this mimicking pattern.

"These findings help us realize that those people who are anxious by nature are the same people who will suffer the greatest harm from sleep deprivation," said Professor Matthew Walker, one of the researchers of the study.

I remember reading Harvard psychologist Daniel Wegner describing this sad spiral in some journal or other a few years back. He gave it a name: Ironic Process Theory. He said that trying to sleep by attempting to eliminate negative thoughts upon hitting the pillow, or trying not to panic about how you haven't slept in three days, or whatever mind control you've been told to try, only succeeds in triggering an internal monitoring process that watches to see if you're succeeding. Which keeps you awake.

I've worked on my insomnia from all angles. And, yes, I've

tried chamomile tea. And sleep meditations. And I know some of you are busting right now to suggest I try melatonin, or counting sheep backward, or earthing mats. To which I would reply that Valium and a pick-axe to the head fail to take the edge off when I'm in one of my sleepless ruts.

But I had some spacious thoughts about it all recently.

When we're babies the mortal terror of the vast, seemingly unsafe experience of life outside the womb is overwhelming and leaves us on high alert and unable to sleep. Our parents must, night after night, hold us tightly and rock us gently, to get us to drift off. Over time, the holding, the rocking, the reassurance of a full tummy, warmth and everything else the baby books advise our parents to do, make us feel safe and supported, and we learn to trust life, and to self-settle. Well, most of us do. Some of us, though, do not learn how to self-settle, or have had reason to unlearn this ability to trust later on down the track (we're abandoned, abused or a younger sibling rocks up and takes away the attention). We feel unsupported and unsafe and so we must remain hyper-vigilant. At night, we simply can't shut down our thoughts and fears. We can't rest easy and trust that the stove is off, that the noisy neighbors will eventually quiet down, that work stresses can be put on a shelf for eight hours. We have to stay "on." We are on our own. We feel this acutely and, oh, it hurts.

Thinking about insomnia this way has helped things a lot. Rather than feeling I have a hopeless, helpless affliction, I can see I just need to find a way to feel held. To feel that everything fits. That everything is going to be okay. That life has this one, my little friend.

I think this is the better journey. What do you think?

13. Here's another equally generous and spacious thought. I throw it into the mix because I'm aware we need lots of different takes to work our way through such a complex condition.

Insomnia is a cry from our core to spend reflective time with ourselves. As British philosopher Alain de Botton puts it, "It's an inarticulate, maddening but ultimately healthy plea released by our core self that we confront the issues we've put off for too long. Insomnia isn't really to do with not being able to sleep; it's about not having given ourselves a chance to think."

De Botton argues that this need to reflect quietly (to reacquaint our selves with ourselves), without the distraction and obligations of our daylight selves, outweighs the benefits of sleep and so we subliminally make the call: think, not sleep.

I come to deep, hurtful but ultimately growth-creating realizations at night. In daylight I struggle to see my true weaknesses. And I wear masks during the day. The eeriness, the loneliness, the expansiveness, the "out-of-sync-ness" of 4am sees me delve into truths and realizations I wouldn't otherwise.

It's *always* 4am, isn't it? Indeed there's a meme doing the interweb rounds that captures all the 4am references in movies, sitcoms and books. There are hundreds. It's a thing!

When I can't sleep now, I remind myself that it might just be about a need to reacquaint my self with my self.

Yep, a better journey.

the something else

14. This is the bit in the book where I go back to my childhood briefly, mostly because my publisher asked me to include it. I've got to say, I do so reluctantly. From years of doing it for counselors and psychiatrists ("And what's your relationship with your mother like?") I find it kind of boring and indulgent. But I relent . . .

I wasn't a morose kid. Although I did scowl in a lot of family photos. I went about normal kid things—collecting smelly erasers, BMX racing, catching freshwater shrimps in our drought-sucked dams.

I grew up in the bush outside Canberra on what my dad described as a semi-self-sufficient property. Not because Mom and Dad were hippies. It wasn't that idyllic, I promise. They were just broke. We bred goats for milk and meat. We lived minimally in gum boots and flip-flops, depending on the weather, and third-hand clothes. My grandfather worked in a factory tearing up clothing that couldn't be sold in the thift stores into rags for car mechanics. He'd set some of it aside for us. He'd arrive at our place with big plastic bags of worn-out rugby league sweaters, hyper-color tees and drop-waisted dresses and we'd pick out our wardrobes.

Mom would buy day-old bread from the Tip Top bakery for 10 cents a loaf. The bread was reserved for pig farmers. We didn't have pigs, but Mom felt justified in the deceit. "I'm not lying. You lot are pigs," she would say.

I hung out with my five younger brothers and sister. We were each other's best mates, in part because there was no one else.

But gravel churned in my brain. Try doing kid things and smiling for photos and being simple and innocent with those little rocks grinding relentlessly. It's distracting.

When I was twelve, two things happened. First, I became obsessed with spirituality. I took to reading the Bible while folded over the end of the bed, blood rushing to my head, running a ruler down the lines. Then I started tapping and counting things over and over and stopped sleeping. I was diagnosed with anxiety for the first time shortly after.

I don't think this was a coincidence.

Church left me deeply distressed. It wasn't fear of God or damnation. It was the deep, deep loneliness that it triggered. At Sunday Mass I would turn around to look at the people behind me. Their calm Sunday morning faces said to me they "got it." *This* is what terrified me—the fact I didn't get it. What was I missing? Mom would tap me on the knee and tell me to stop staring. The despair crept over my shoulders. I associated church with the too-tight-around-the-neck prickly Fair Isle sweater I wore at this age.

When I was about thirteen, I declared I'd renounced God, having looked into the matter at length. I subjected my parents to explosive rants arguing that church was really rather detrimental to my well-being. I quoted the Bible, mostly the

Old Testament (I never did read as far as the new one in my research), to make my point.

Somehow I convinced Mom and Dad that I had to try other religions, to see if they "fitted." Exhaustion and bewilderment saw my parents give in. Which was unusual. Normally my anxiety and panic was something they felt they had to quash. It had to be contained, toned down. It was too much. I was too much. I appreciate the burden they must have endured now. But back then I was all raw exposed nerves with those damn gravelly thoughts rolling around in my brain and nowhere to place them. No one to steer me or hold me.

And so each Sunday we'd drive the hour or so into town and they'd drop me off at different temples and conference rooms that I directed them to, while they all went to the Catholic church with the post–Vatican II beer-bottle windows and utilitarian wood veneer pews.

I don't have clear recollections of this early teen spiritual search. I don't know how I found these alternate places of worship, pre-internet. I asked Mom. She doesn't know either. I can only assume I researched using the Yellow Pages. I know I loved to read the Yellow Pages, and junk mail. It was incredibly soothing to compare prices. To compare anything. And count things. In sets of threes and multiples thereof.

I have glimpses of sitting in a room above a Chinese restaurant with a scientologist who held a clipboard and wore a khaki shirt tucked into beige pants. I recall soon after sitting in a plastic stackable chair at the back of a Hare Krishna meeting next to a fold-out card table holding pamphlets. The people in front—men and women—chanted and clapped, which made me awkward and overwhelmed. When Mom and Dad picked me up two hours later I was sitting on the nature strip out front crying, in an anxious panic. I *still* didn't get it.

I also don't have much memory of going to the counselor who told me I was anxious. Actually, this is a common theme across all of my anxious episodes—very hazy recollections of everyday details, like dates and locations, but sharp-as-tack memories of the theories and ruminations that accompanied the episodes. Anyone else experience the same?

I remember walking in through glass sliding doors and down an asymmetrical corridor with craft projects hanging from the ceiling beams—paper mobiles and finger-painted things attached to string. I know I was seeing a counselor to address the fact that I wasn't sleeping. But also because of the tapping and counting. There was an ankle-grabbing beast under my bed that I knew would grab me if I had to get up to go to the bathroom (which I always had to). The combination of fears left me paralyzed. I'd have to lean over and check to see if it was there, for hours on end and to a count of three. Or four. Depending.

Mom and Dad thought I was bored. "Uh-oh, Sarah's bored again," they'd say if I got myself wound up. It was one of those things that was said in our family. I wasn't bored, though. I was frantically trying to get answers and flee the fluttery-ness in my guts.

I was running around with a hot potato with nowhere to drop it off. I got even more anxious when I became aware that no one else seemed to be feeling the same things. I'd lie awake all night at sleepovers hyperaware of everything about my friend's existence and not knowing where to place all the thoughts and analysis. *Wow, Penny's mom doesn't wear a bra, what does that mean? Apricot chicken from a can . . . is that allowed? What do other people think about when they're falling asleep?*

I also set up a business about this time. And this, too, was about fleeing the fluttery-ness. That's how I interpret it now. When I say "my business," let's be frank, I was twelve.

Or thirteen. I was a lackluster artist but I would spend weekends making library bags from a roll of calico that I then painted with dinky pictures of sunbaking elephants and galahs in aprons. I also did a range of Australiana brooches, and hand-painted gift cards sold in packs of five. I negotiated with Mom to drop me off in town once a fortnight or so while she picked up supplies on one of her weekly trips in from the bush (they could only afford petrol for one trip a week) and I'd front up to boutique toyshops and galleries with my wares. I'd like to say I was a passionate craftsperson. But really it was more that making stuff and having an enterprising meta-purpose distracted me while I tried to grapple with the meta-issue of my existence.

Georgia, one of my editors, has made a note asking if the business was successful. I guess you'd have to say it was, G. I know I made $7 a bag. — Dad used to call me "the little capitalist" and I was the family loan facility.

I point out this nexus—between my spiritual yearning and anxiety—because it's helped me understand my restlessness ever since. Anxiety and existential curiosity are connected. Yes, absolutely, it can become medical when it spirals out too far. But its origins are far more fundamental.

15. In my early thirties I went on a yoga retreat. It was before what I shall refer to as my Mid-thirties Meltdown. I was in the middle of a messy, very protracted breakup. I was editing the Australian edition of *Cosmopolitan* magazine at the time and I'd started to lose my focused, bird's-eye grip on my staff, on my ideas, on my principles. I was looking for something, even just a break in the rut in my head.

I've done this since my early teens—revisited my spirituality on and off. And it's always occurred in tandem with my anxiety. Indeed, I enlisted the help of a spiritual counselor during this time. I saw Sky, a tall and magnificently elegant woman in her late fifties who'd lived a very full life, on Tuesday afternoons and charged her with "keeping me real" amid all the handbags and free eye creams and client cocktail dinners.

I arrived late for the three-day vinyasa intensive at the collection of wooden huts tucked into dense bushland a few hours south of Sydney. I do these retreats, but I tell you, every single time I arrive fidgety and cynical and worried about sharing bunk beds with strangers and doing partner-up-with-someone exercises.

And constipated. I always, *always* get constipated at these things. I've since asked around—lots of other A-types do, too. I pass through the gates of the retreat center and everything just jams right up. Sure, I come across all divulge-y and confessional to most people around me. But I'm like the kid who's told to share his bag of lollies and dutifully holds out his stash, gripping it from below so that only one or two he doesn't mind giving away can be picked out.

You don't have to be Louise Hay to see what this is about. I hold on to my crap when truly confronted. Funnily, always, without fail, on the last day as we do our final class or meditation, I finally let go.

I poo.

Then I cry.

Then I drive home.

On day three of this particular retreat, we were sitting in an old scout hall, the doors flung open to bushland. It was late afternoon and we had to meditate for a full hour. I'd never sat still so long, nor properly meditated. I gripped. I wanted to run. To climb a tree. Couldn't we just *talk* about meditating in expansive ways over rooibos tea from the urn in the common room? But I sat and followed the instructor's directives. It went a little like this:

SIT ON A SMALL BENCH WITH YOURSELF

Sit. Quietly. Turn your awareness to your heart space.

Now imagine you're sitting on a small wooden bench with yourself. Imagine you're doing so in that space in the center of your chest. There you are, sitting to your right, the little nattering humanoid that you are, berating yourself for eating too much at lunch and debating whether to hang the washing out or not. This little nattering self is your little "i." You (the big "I") can watch it all. Yep, there you are, sitting quietly, looking out at a view, over treetops down to an ocean. On your little bench. Together. You're just hanging, nowhere to go, nothing to do. The two of you.

You will probably pull back from the heart space into the head to analyze and commentate. And to grasp outward. Because that's what we do. What will I eat for dinner? Are everyone else's legs falling asleep? But when this happens you gently swing your internal, closed-eyed gaze back to the situation on the bench. Yep, there she is, your little mate—your little "i"—to your right.

And you hang out a bit more, calmly and patiently.

And then it might occur to you that your little mate "i" is just that—a little mate sitting next to you. And that this Big "I" is who you really are. It feels deep and close and yet so vast.

I find this realization funny when it strikes. Here "I" am. And, yet, for so long I thought I was merely my little nattering mate sitting next to me. It's funny in the way that a kid finds it softly, self-consciously funny when they realize their dad has played a trick on them. A dawning funny. I realize in this moment that I have been here all along.

I sit in this. It feels like when I climb into my own bed after being away traveling for a week. The smells and the warm hug at my being. It's also spacious and expansive, open-chested to the world. I feel spacious. I feel the world is spacious. It's magnificent and elevated.

I don't want to run from it like I normally do, back to my scheduling and thinking about the load of washing I'll need to put on when I get home. Or, worse, to my inner self-berating. *Sarah, everyone else is doing this just fine; find a way! Sarah, you should be doing more productive things. QUICK, you're wasting time.*

16. I hear the birds outside doing their late afternoon restless thing, furtively dashing about like it's department store closing time. This fluttery, end-of-the-day sound can often get me fluttery, too. But something's different this afternoon. There's a sense of resignation, and I go along with it.

The birds descend into the tree outside, one by one, quieting. A few swoop up and about, causing a cluster to loudly bustle up in fright. I know that unsettledness. It's what "i" do. But I hold the feeling and come in closer, on that bench, to my deep, vast self. And gently, gently, with a little more holding, we all settle. We land.

Then I poo.

Then I drive home.

I'd never been able to meditate previously. Or at least not comfortably. I'd always been too anxious. I'd sought out meditation many times. I had dropped into community classes and went on one of Sky's retreats. I tried to grasp that spiritual connect I was looking for, but it never fully clicked. When I was eighteen I organized a meditation course for anxious women on my university campus. I brought in an instructor to lead things. I sat in the group and joined in. For thirty minutes I effectively sat cross-legged and trash-talked myself. It was brutal. At the end, everyone walked out peacefully, gently sharing their relaxing experiences. Me, I fled to the toilets in the student politics common room and bit into a roll of toilet paper to hold back a long, guttural scream. It was too much of a gear change back then and the half an hour of desperately trying to put the brakes on my frantic overthinking only made things worse. So did the acute awareness that everyone else seemed to be able to do something that I couldn't.

But in that moment in the scout hall with the sun streaming in, I touched a still, settled, vast, spacious, magnificent knowing at my core. It was only for a few delicate moments, but there was no going back. The scab was removed and the rawness—the "Something Else" I'd been looking for—was finally exposed. I call it the Something Else because there's no other way of describing this yearning—this indescribable thing or place or energy I'd been looking for—that came before words.

But now I'd touched it.

And goddamn I wanted to touch it again.

17. Have you read Dodie Smith's *I Capture the Castle*? Bookish friends tell me they read it as kids. Reading it now in my forties I don't see it as a kids' book. I've drawn from it all kinds of grown-up spiritual themes. The protagonist, Cassandra, a sharp-minded teen seeking the Something Else, asks the local priest, "And do religious people find out what it's all about?

Do they really get the answer to the riddle?" The priest replies, "They just get a whiff of an answer sometimes . . . If one ever has any luck, one will know with all one's senses—and none of them. Probably as good a way as any of describing it is that we shall 'come over all queer.' "

"But haven't you?" asks Cassandra (which reminded me of the same question I asked the Dalai Lama in relation to quieting the mind).

The priest sighs and says the whiffs are few and far between. "But the memory of them everlasting."

18. I'm hoping you know what I'm talking about when I talk of yearnings and ephemeral Something Elses and whiffs and things. Maybe this will help:

You know how dogs do that thing where they circle and circle, unable to find the spot where they feel comfortable enough to settle? That's us. Most of the time. We wander about, filling up our weekends, creating never-ending to-do lists.

It's like we're searching for a Something Else that makes us feel . . . what? Like we've landed, I suppose. And that things are all good on this patch. Know what I mean?

And so here is my (possibly) contentious theory, the one that has guided me for some time now and that I took to my publisher when she asked me if I'd write about my anxiety:

Anxiety *is* a disconnection with this Something Else. As I say, the doctors and scientists can call it all kinds of things, but I believe it all comes down to this disconnect.

As German psychiatrist Karl Jaspers wrote, anxiety is a feeling that you've "not finished something . . . that one has to look for something or . . . come into the clear about something." To put it bluntly, as Jean-Paul Sartre did at some point, unlike a table or garlic crusher or whatever, which has a comfortingly obvious purpose, we don't know why we exist.

This, dear friends, is what I am talking about. Yes, yes, yes. This is anxiety.

It's this lack of connection and clarity that leaves us fretting and checking and spinning around in our heads and needing to compensate with irrational, painful behaviors, whether it be OCD, phobias or panic attacks. It's this sense of *missing*... something... that leaves us feeling lonely and incomplete and fluttery. Something is not right. We haven't landed.

When I fret that someone hasn't rung me when they say they will, for example, sure, I'm fretting about feeling abandoned and lonely in life. But let's pare this back another layer. I'm really fretting that I'm not able to exist calmly, happily on my own, on my own bench.

When I check that a tap is turned off, perhaps thirty-two times, in repeated patterns of 16 x 4 sets, between 2am and 5am, sure it signals trust issues. But drilled down, I'm really fretting that *something* is missing that should be making me feel supported, comforted and assured that everything's going to be okay. That I'm not connected with this Something Else *is* my anxiety.

I checked in, once again, with a number of psychiatrists on this rather contentious idea, too. Was I on a dangerous (possibly manic) tangent? SANE's Dr. Cross tells me I'm not when we meet for coffee at my local café-slash-office. "Anxiety is *all* about a lack of connection and a need for spiritual answers," he says. He even directs some of his patients to a "spiritualist church" in Sydney's western suburbs where existentialist topics are discussed. After we finish chatting, a couple at the neighboring table lean over and explain that they are psychologists and are about to publish a neuroscience paper looking at how the chemical imbalance model was failing anxious patients and that they couldn't help but interrupt to say they agreed. Margot and Dale are research partners and it turns out that I know of Dale's previous establishment-stirring work (he published the bestselling *Don't Diet* in the late '80s). Margot says, "For hundreds of thousands of years our ancestors have asked the same questions: 'Who are we? Why are we here? Is there any point?' Why do we think we should suddenly block ourselves from asking such questions now . . . which is what the current medical model does." She apologizes for interrupting. No, no! It's perfect, I tell her.

So let me be really firm on this—I believe the yearning for this Something Else buzzes at our cores. It's the soundtrack to our lives as we go through the motions of doing our tax returns and walking to the pharmacy to buy aspirin. I talk more about this later, but for now let's hand it to Buddhist monk and author of *The Miracle of Mindfulness* Thich Nhat Hanh for the final word:

> *You want to find something, but you don't know what to search for. In everyone there's a continuous desire and expectation; deep inside, you still expect something better to happen. That is why you check your email many times a day.*

19. Although, granted, some of us are able to put on noise-canceling headphones. And, about three times in any given week, I envy such people this most sweet aptitude.

I call these people life naturals. In fact new research shows that 20 percent of us (well, not me personally) were born this way—biologically immune to anxiety. Turns out they have a very fortunate gene mutation that sees them produce higher levels of anandamide—the so-called bliss molecule, similar in effect to marijuana. Which might explain some people's inability to arrive anywhere on time.

I've always dated life naturals. Life naturals can be a salve for neurotics.

My Mom is a life natural. She sleeps soundly at night and can tune out of the chaotic, competitive conversations at the dinner table at Christmas when it's the best thing for her well-being. She'll sit there with a quiet smile as we all bait each other, referencing gags from thirty years ago, bagging out Dad for listening to the extended version of "The Only Way Is Up" by Yazz while he does the vacuuming. Dad, on the other hand, experiences the buzz at full throttle. "Well, with your father I didn't sign up for the merry-go-round," she said to me one winter morning sitting in their kitchen as Dad got himself worked up about how he'd fit in a two-hour run between the dozen or so other tasks he'd assigned himself for the day; "I signed up for the roller coaster." I guess roller coasters can be entertaining.

This is what else life naturals do: they see a flower. And find it beautiful. That's it.

They don't wonder if they're liking it enough, or if the whole experience is a waste because today they're too stressed to appreciate lovely things like flowers. Nor do they fear that the flower won't last. And they don't try to draw on that Zen proverb about how a flower doesn't try to bloom, it just blooms on its own. And then despair that they're failing to do the same. They simply grasp the is-ness as a matter of course.

This is another thing life naturals do: they can see straight to the positive. Here's an example. I was with the guy I dated while writing this book, reading the paper after a very fresh ocean swim. We'd met on Tinder. He's an electrician and fisherman and has lived on the Northern Beaches of Sydney all his life. He's a caricature of laid-back, even for the Northern Beaches, an area renowned for its she'll-be-right-while-ever-the-surf's-up culture among blokes. He's a Life Natural.

I watch as four Maseratis pulled up outside a café at Sydney's Palm Beach, revving obnoxiously for no reason. A bunch of middle-aged blokes climb out of the cars. One of them *actually* has a lemon sweater knotted casually around his shoulders. It's a lot of things all at once. My cynical head rears like a dog to a cat and I grapple for the most cutting summation of all the vulnerable, totally human stuff going on in this scene. "Do they *truly* think that we're thinking anything but 'terrible cliché?' " I ask.

"I don't know," says the Life Natural. "I just reckon they're happy they can afford the car they've always dreamed of owning." Yeah, alright. Fair and generous point, I told him.

Anyway, this book is not for life naturals. Unless they want to read it to better understand the roller-coaster ride they've chosen to be on. Which is actually not a bad idea.

gentle and small

20. For the past six years, when distressed sugar quitters have contacted me worried that they've "fallen off the wagon" and that, henceforth, their entire life is going to fall to pieces, I tell them to eat a pork chop. With some sweet potato. And some steamed zucchini doused in olive oil.

The anti-sugar crusade that I'm told I'm largely responsible for, and that now sees me travel the world for five months of the year, publishing books and running a business with twenty-three staff from a converted warehouse (with a worm farm on the rooftop balcony and all manner of clichéd internet start-up accoutrements), began not long after my anxiety flared in my Mid-thirties Meltdown and I fled to an army shed on seven acres of forest in the Byron Bay hinterland. Because that's what you'd do, right, when the proverbial hits a fan? Some might say this move marked the turning of a new leaf. I wouldn't. The leaves have never stopped turning.

I made the decision to up and leave and within a week I was on the road with what I could fit in the back of my Holden

Astra hatchback: a slow cooker, a blender, a few changes of clothing, basic toiletries and my bike. I gave everything else away at a garage sale where you could name your price, the money going to the public school up the road. It was easier than haggling. I didn't want to bargain or hustle, or deal with humans IRL (in real life) for a long while. I stayed in my shed, alone, for eighteen months, supporting myself by writing that magazine column I mentioned where I investigated ways to live a better, more well life.

The shed was one room with a potbelly stove in the middle. You climbed outdoor stairs to a bedroom in the roof and the bathroom was outside, too. From the bath I had a view of every moon that the cashed-up kids with their unicorn tatts down on the beach would bang drums to each month. There were no locks and lots of bush turkeys.

One week I was short of a topic and so I quit sugar. It was all very convenient. I'd been told I should quit my seductively gnarly habit by several doctors and naturopaths; an editorial deadline provided the impetus. Seductively gnarly? I was one of those types who ate "healthy sugar"—honey in my chai tea, dates and banana on my maple-frosted granola, gluten-free muffins, and so on. I convinced myself, and everyone around me, that I didn't have a problem. I was, in fact, eating almost 30 teaspoons a day. It's never surprised me that sugar addiction goes hand in hand with anxiety, and that anxious folk hide the vice so protectively. We're dopamine junkies, and we don't like people removing our "fix."

That was January 2011. I tried the no-sugar thing for two weeks. It felt good—I felt calmer and my skin glowed—so I kept going and going. The experiment turned into a series of blog posts in which I shared my research and recipes, which turned into an ebook (I taught myself how to make such a thing, squirreled away in my shed), which turned into a bunch of print books, which turned into an online cooking program, which has seen two million people ditch the white stuff.

Anyway . . . when I put together the program, I was heavily influenced by the Buddhist notion of compassion.

You might have done the 8-Week Program. If so, you'd know how much I bang on about being "gentle and kind" in the process.

Draconian, self-flagellating diets don't work. This is because they're anchored in the *not doing* of something. We humans, I feel, are far better at *doing* something. Show me a "Wet Paint: Do Not Touch" sign and all I want to do is touch it. So to make change, it's best to ditch the self-flagellation. I *invite* people to simply give quitting sugar a go. No big deal. I suggest you just commit to two weeks and see how it goes. "Suck it and see," I say to really back the pressure off. I focus on all the abundant food you *can* eat (*haloumi cheese! red wine! macadamia nuts! butter! cauliflower pizza!*), rather than ranting on about "bad" foods. I work to an eight-week timeframe that allows for lapses because the latest addiction theory shows that allowing sixty days is more effective than the previous twenty-one-day edict, which didn't leave any room for "failure." If you fell off the wagon, back to the beginning you'd go.

So, as I tell readers when I proffer my pork chop solution, a sugary lapse is not a problem. In fact, it's a good thing: the resulting frazzled, unstable, foggy feeling reinforces why you quit in the first place. And why the pork chop? From a nutrition perspective, porcine protein is a boon for recalibrating your appetite hormones and blood sugar levels, a theory that pivots from work I did with the *National Geographic* team that studied the lifestyle factors common to the longest-lived populations around the world (hearty ingestion of pork and red wine are among them).

Plus, the notion of solving a problem with such comforting food tapped into the gentle and kind vibe I was feeling so strongly about.

While all of this was happening the treatment of anxiety was undergoing a massive shift, one that I followed and reflected on via my personal sarahwilson.com blog. It was getting gentler and kinder. If you're someone who's taken even a few steps on the anxious journey any time since the late 1990s, you've probably encountered a bunch of CBTs and been bludgeoned with various positive psychology and self-improvement theories. Apart from being gratingly annoying (*sand play! yellow smiley faces! Bobby McFerrin!*), these approaches were about changing yourself and reversing your behaviors (*turn that frown upside down!*). As a result, I and many others developed a visceral wariness of them.

But now, CBT and NLP are being surpassed by acceptance and commitment therapy (which focuses on accepting your thoughts and emotions and living to your core values) and compassion-focused therapy (which incorporates Buddhist mindfulness practices to increase your psychological flexibility). Similarly, positive psychology has mercifully given way to what's being referred to as a "second wave," and, instead of being told to have a "Happiness Makeover" and to "Change Your Thinking," we're being asked to "Lament the Loss of Sadness" and to "Go Against Happiness, In Praise of Sadness" and get on with "Embracing the Dark Side of Life." And, most recently, to try out the "The Subtle Art of Not Giving a F*ck" (the title of a book by U.S. marketeer Mark Manson, who advocates giving up on trying to sidestep pain).

These approaches are rooted in working with what "is" and easing our way into the life we want, gently, kindly. Instead of building a bridge (with happy-clappy language and unicorn emoticons) and getting over it, we make the most of the river we find ourselves in, even if it might be a little dank and overgrown with reeds at times. By doing so we may find happiness, among other different, rich emotions available to us. Happiness is a lovely by-product of the process. Not the (mostly unattainable) end goal.

Ruth Whippman, author of *The Pursuit of Happiness and Why It's Making Us Anxious*, reckons the search for happiness is making anxiety worse because "the expectations of how happy you should be are so high, you always feel you are falling short." Whippman, somewhat acerbically, argues that our pursuit of happiness—including the recently fashionable route via mindfulness—is particularly privileged. "It is a philosophy likely to be more rewarding for those whose lives contain more privileged moments than grinding, humiliating or exhausting ones. Those for whom a given moment is more likely to be 'sun-dappled yoga pose' than 'hour 11 manning the deep-fat fryer.' "

(I've previously wondered if it's a privilege to even be able to question your anxiety as I do here—whether anxiety is a bourgeois affliction. I don't anymore. I think the issue has gone rogue—more on this to come—and while not everyone can book themselves into a meditation retreat in Sri Lanka for two weeks at a time, or flee to a shed in the Byron Bay hinterland, most of us really do benefit from knowing we're not alone and from trying out each other's simple techniques for accessing some stillness and peace. But I digress . . .)

And here's another issue with the "just be happy" approach of the past fifteen years: happiness is put forward as a choice, not a matter of luck. Yet happiness literally derives from the Middle English word *hap*, meaning chance or good luck (thus "happenstance" or "perhaps"). We've twisted the meaning in recent times such that it's now something we just have to work hard to get to the bottom of. As though it's an endpoint that exists. We just have to sift through various options and decisions and choices. But, of course, getting to the bottom of options is anxiety-inducing. Whippman refers to stacks of studies that show that the more relentlessly we value and pursue happiness, the more likely we are to be depressed, anxious and lonely.

Up in my army shed in the forest, I was learning to apply kindness and gentleness to my health. I was too exhausted to flog myself with restrictive eating or hardcore exercise. And all the wisdom I was learning from my research reinforced the efficacy of this softer way. So did the online conversations I was having with all the vulnerable sugar-addled strangers out there who I'd come to feel so deeply for. We were all wanting a kinder, gentler way. We were all tired and sad.

Wonderfully, perfectly, this kindness and gentleness seeped into my anxious journey, too.

21. It is not our fault that we're addicted to sugar and getting fatter. Obesity is not an issue of self-control, or lack thereof. Every reasonable and respected voice in the debate now acknowledges that the way in which we're force-fed sugar by the food giants deranges our metabolisms to such an extent that for many no amount of willpower, hardcore workouts or dietary commitment can make a difference. Children today are being *born* with type 2 diabetes. Low-fat "plain" yogurt can contain up to 6 teaspoons of sugar per serving. We can no longer point the finger at individuals caught up in the unfortunate metabolic cycle.

Just as we can't blame those of us with a highly sensitive amygdala for being anxious.

22. While writing this book, I read *The Compassionate-Mind Guide to Overcoming Anxiety* by cognitive therapist Dennis D. Tirch, a founding voice in the compassion-focused therapy movement. Tirch tells us that our fight-or-flight threat detection system evolved as an essential part of our survival and that some of us are born with more sensitive detection switches than others. It's not our fault if we drew this straw; it's simply where the evolutionary tide landed us.

And, he adds, even our best attempts to avoid or combat or criticize our anxiety will only make it worse. Instead, self-compassion is the way forward.

Tirch and others explain that we have an old brain (the one that evolved first) and a new brain. The former takes information drawn from our senses, and when it perceives danger it stimulates a very "old" gnarly part of the brain called the amygdala, which activates the stress response, releasing the stress hormones adrenaline and cortisol. In an instant we're ready for action—to fight or to flee. This part of the brain doesn't distinguish between real and perceived threat. It just reacts, fast.

Our "new" brain, crudely put, imagines, plans, thinks and is responsible for our self-awareness and our sense of mortality. It's the bit that defines us as humans. We anxious folk have particularly active new-brain stuff going on. We think a lot. We're extremely self-aware.

Our old and new brains collide frequently, and problematically, so if your new brain is particularly sensitive, as they are for anxious folk, your sensitive reflections and thoughts may trigger your gnarly old "threat" system into believing a saber-toothed tiger is about to pounce, with all the accompanying hormone surges and their physical side effects.

But you might be interested to know that in addition to the threat system, Tirch explains that we have another protective mechanism—the comfort system. As a species without horns or venom, we've relied on strong communal bonds to survive, bonds that are created and reinforced with soothing and comforting behaviors. To survive as children we needed our parents to comfort and soothe us as we learned skills and developed strength. This system activates the feel-good hormones oxytocin and endorphin, which effectively shut down the threat system.

Like the fight-or-flight response, the comfort system is also automatic and will do its job in toning down anxiety . . . if we learn to trigger it. Studies show that one of the best ways to trigger the comfort system is to practice self-compassion. Which is all very well in theory. I don't know about you, but when I'm in one of my self-loathing whirly-whirlies I'm not feeling a great deal of soft-eyed, open-armed love for myself. Fortunately the new-wave behavioralists have this one covered. They acknowledge that it's easier for self-flagellators like myself to activate compassion for another than it is to activate it for ourselves and conveniently supply studies that have found showing compassion for others will have the same comfort system activating response in the brain, thus dampening the anxiety-riddled threat system.

Of course it's another theory in a long lineup. But it's one that makes intuitive sense to me. It has for a while.

TALK TO A KID

Here's a good way to trick yourself into some self-compassion when you tend to be more of a self-loathing type.

I'm not a big fan of imagination exercises. But this one is effortless and switches me into the right gear really quickly.

I don't plan on filling this book with too many exercises that people like me skim over to get to the meaty theory. I'll just include the ones that suit people like me and you.

Picture a child, preferably you as a child, or another child in your life you rather like. This kid is upset and anxious because they haven't finished their homework, or they're distressed about being picked on, or they're feeling lonely or abandoned or confused. Then imagine the compassion you'd feel for them. Now tell the little kid this: that you get it, that you can see what's going on. Tell them they can't be blamed for feeling as they do, and that they won't feel this way forever (or whatever it is you feel for this funny little lost child in your mind's eye).

I dug up a kindergarten photo of myself not so long ago. Having a photo in front of me helped me do this exercise. In the picture I'm wearing my hair in pigtails and I have a scab under my lip where my bottom teeth had cut all the way through when I fell down my grandmother's stairs. I still bite the scar when I'm nervous today. I see in my eyes the most intense earnestness. I remember the photo shoot. I remember wondering how to smile for the photographer. I'm able to feel compassion for the mini-me. My heart goes out to her. I want to hold her while she works it all out (while not denying her the process of doing so). I tell her that she's all good, that she's on the right track. That she's a really cool, interesting kid for wondering about all these things and that as she gets older it's going to make her an amazing person to be friends with, and an excellent big sister. I tell her I'll be there with her the whole way, so she has nothing to worry about. When I'm done I just sit in this teary-eyed feeling for a bit.

WRITE A "NO BLOODY WONDER" LETTER TO YOUR ANXIETY

This is another fool-yourself-into-self-compassion-and-thus-activate-your-comfort-valve exercise. I like it because it connects me to my anxiety and gets me real about what's going on. I start, *Dear Anxiety, you funny little thing*... I like seeing my anxiety as cute and endearing in its earnestness, much like the kid at the sport carnival who pushes super hard in the 200-meter race, all red in the face, about to explode. I go on to acknowledge what it's up to, what it's feeling. I continue with "No bloody wonder..." and validate why it's got itself worked up. *No bloody wonder you're wobbly—you've been left in limbo for three days over a work outcome yet again. Plus you feel like you're in a rut, unable to get a clear view of why you're living.*

Then I might get just a little bit "walk down the hall of mirrors" with it. *Yes, yes, I know it feels like it's too hard. But you deal this up every time we land here. Let's just look back on it all for twenty-seven seconds. The shittiest days have always led somewhere. Haven't they?*

Last week we fretted all Saturday morning and it was a glorious day and we got paralyzed on the lounge-room floor and it was all such a waste of a glorious day. I know. I was there. And the fretting got worse and tighter. Until we cried. And it all felt good. And we realized we hadn't cried for the bigness of life for too long. We had to fret to cry to release the pressure.

So let's just sit in this for a bit and see what happens. Let's be totally grim together. Might as well. We're here anyway. Cool. Okay then.

As I do this I'm also able to offer my funny little anxiety some helpful ways forward. *You know what? I think you just need to go for a hike this weekend. Get into the bush.* As I sign off, *Love, Sarah,* I smile at my anxiety.

23. When I was twenty-three, a ninety-two-year-old Russian-Chinese hypnotist introduced me to the idea of mental muscle building.

Eugene Veshner was a former civil engineer who'd been told he'd die at forty from a congenital disease; he'd already turned partially blind. To deal with the pain of such news he used his scientific brain to develop what became an internationally recognized self-hypnosis trick to shift his outlook, which, to everyone's surprise, saw the guy keep on living and living. And regain much of his eyesight.

I was Eugene's last patient. He'd retired the year before, but he took me on because, in his softly spoken words, "You're messy." I'd just returned from the U.S. where I'd been studying on a scholarship. But I'd had to come home early due to illness. It was a perfect storm of manic depression (as it was still being called back then) on top of the obsessive-compulsive disorder, insomnia and a roaring case of Grave's disease, an autoimmune disease of the thyroid that preceded the Hashimoto's disease I developed later, in my thirties. I'd been reduced to a size 6 from a size 12 and I gasped for breath.

"Let's start with the insomnia," he suggested.

I loved Eugene. I loved visiting his modest brick veneer home liberally snowflaked with doilies and little vases of violets. I loved his woolen vests and his long fingers. "Bad habits [and insomnia, he told me, is a bad habit] can't be reversed or eliminated. It's not how the brain works," he explained. He drew a line on his notepad with his fountain pen. "This is a habit, a series of thoughts. They clump together to form a neural pathway and the more thoughts you add to this the thicker it gets." He draws more lines over the top of the first.

"You don't *delete* a bad habit, you *build* a new, better one. You feed this new habit, over and over," he tells me. He draws a new line, this time parallel to the first clump of lines, and thickens it with more and more strokes of his pen. The new thoughts clump, layer by layer, and eventually create a habit that is stronger than the old one. You build habits that trigger the comfort system, instead of the threat system.

My old habit was thinking I had to check under the bed for something nebulously dangerous (mostly that ankle-grabbing beast) . . . again and again. Which obviously strengthens all that fighty, flighty stuff. My new habit was getting the urge, and resisting it calmly. I visualized this in a calm, meditative state of self-hypnosis, the best state for drawing new lines. I pictured lying in bed and being cool with not checking. I created this picture, over and over, in my mind. After about three weeks of doing this every night, it played out in real life, in bed one night. I went to bed and I lied there. I reproduced the calm of the imagined scenario. I stayed. I stayed. I kept breathing. I was aware of the visceral urge to check. But I stayed. To see what happened. As I waited, I drifted off to sleep. In the morning I grabbed one hand in the other and shook myself in congratulation. My goodness I was proud of myself.

Slowly, patiently, I've worked through years of anxious habits like this, one clumpy neural pathway at a time, strengthening my comfort system. It wasn't about changing myself. It was about creating ease and gentleness around who I was, which allowed me to make better choices.

Many say it takes sixty-six days of *continuous* work to create a new habit (I like to think of it as building a new muscle). To be any good at anything, whether it's archery or writing novels, they say it takes 10,000 hours of work, or building mental muscle. I'd suggest you triple that if you're wanting to manage your anxiety effectively.

Heck, make it a lifetime.

Neuroscience now tells us that our brains are not rigid entities. They're more like plastic (or a muscle!), and we can reshape them with small movements over a period of time. Anyone who's read Oliver Sacks, or any neuroscientific work, knows that with continuous practice you can reshape, or re-wire, your neural pathways according to the habit you want to form. Pathways that are not used weaken and wear off in due time, making room for new neural pathways to form. If you don't use it, you lose it. This is why it is easier to form a new habit than maintain an old one.

When I look back on my bumpy journey I'm forced to note that, underwhelmingly, it's been made up of a hodgepodge of simple small tricks which have slowly—oh so slowly—added up to a new way. I share most of them in this book.

Prefer a faster fix? Oh, goodness I did, too. I still do a lot of the time. But to the "me" that hangs on to this notion, I say GET OVER YOURSELF.

MAKE YOUR BED. EVERY DAY.

This is what I mean by small and underwhelming. I've chatted to *The Happiness Project*'s Gretchen Rubin a few times on this journey. Her book had hung around in the *New York Times* bestseller list for more than two years after it was published in 2009 and we'd reach out to each other every now and then to share ideas on how to make life better. I rang her one day to ask for the stupidly simple trick she employs to quash her anxiety. She was previously a New York lawyer. She knows anxiety.

This was her tip: "Make your bed. Every day."

It struck me as ludicrous. Like "42" being the Ultimate Answer to Everything.

Gretchen explained that such simple, outer order creates inner calm. But it was the "every day" bit of her edict that she wanted to stress.

"It's easier to do something every day, without exception, than to do something 'most days,' " she said. "When you say 'I'll walk four days a week,' you debate which four days, and wake up debating whether you can skip Tuesday."

True. True. It sets us up for decision overload. And so we balk. And don't do any exercise and have unkempt beds for weeks.

This trick, of course, activates the comfort system. Over and over.

I'm not a bed maker. But I made mine for over a year after chatting to Gretchen. Not because I see intrinsic worth in having a neat bed (I prefer to air mine with the duvet pulled back during the day and make it up just before sleeping). But because it was building something.

And in case you're left in any doubt, there's always this from Andy Warhol: "Either *once only*, or *every day*. If you do something once it's exciting, and if you do it every day it's exciting. But if you do it . . . almost every day, it's not good any more."

A SINGULAR PAGE ON BEING ANXIOUS ABOUT BEING ANXIOUS ABOUT BEING ANXIOUS . . .

I'll probably flag this a few times. But let's raise it here nice and boldly on a clean page. One of the worst things we can do to ourselves on the anxious journey is to get anxious about being anxious. I think that a good, ooooh, 80 percent of my anxiety comes from being anxious about being anxious. And 80 percent of that secondary anxiety is compounded by being anxious-slash-pissed off that I'm anxious about being anxious. And on it goes compounding on itself ad infinitum. It's a peculiarity of being human; we are the only animals on the planet capable of being aware we're anxious. Our non-human friends just do the fear, or anxiety, and it stops there.

So, one gentle and small thing we can do (actually it's pretty big and fundamental) is to work on stopping the anxious-about-being-anxious cycle. Franklin Roosevelt proclaimed there is nothing to fear but fear itself. I'm kind of saying the inverse. Don't fear the fear. Instead, see it for what it is. You're feeling anxious. You just are. No need to berate yourself for this; it will only make you more anxious. No need to think that things should be otherwise and that you've got it all wrong somehow. For this, too, will just make you more anxious. Maybe your hormones are out. Maybe the wind direction is all wrong. Maybe things are actually quite crappy right now. And, yeah, that presentation you have to give in two weeks is absolutely anxiety-inducing. Got it. But let's just leave it there, and not fret that you're fretting. Yeah?

Because we're human we have to watch ourselves go through this pain. But because we're human we can also *choose* to watch it and see it for what it is. It's anxiety, for sure. But it doesn't have to be a catastrophe with no endpoint. Even a panic *attack* only lasts 10–30 minutes.

Do the anxiety. Then leave it there. This is our challenge.

JUST SAY IT: "I'M ANXIOUS"

This sounds dumb, I know, but just saying, "I'm anxious" can make you less anxious. A recent study used functional magnetic resonance imaging (fMRI) to map the electrical activity in the brains of subjects who were shown a series of pictures of people with different emotional facial expressions. When subjects were shown the pictures only, their amygdalae were activated. But when they were shown the pictures and asked to name the emotions being expressed, the prefrontal cortex was activated and the activity in the amygdala reduced. Also note, the researchers found the less words used the better. "I'm anxious" should do the trick.

This is something that comes up a lot: our moldy old amygdala, which steers anxiety, struggles to fire up – when other stuff is going on. It's good to know; it means we can take advantage of such limitations.

DO CORE EXERCISES

A study published in 2016 found that the area in the primary motor cortex linked to the axial body muscles (our core) is directly connected to the adrenal glands. Work your core, decrease the stress response. Yoga, Pilates, planking . . . it will all build the right muscles—physical and mental.

It makes sense when you think that we've always known the inverse: that a stooped posture from poor core strength is a sign of angst in a person.

just meditate

24. I'll say it dead straight, because this is how it was presented to me: when you're an anxious type, meditation is non-negotiable.

Sky, the spiritual counselor I employed during my time working in magazines, would tell me, "Sarah, just meditate." On my Tuesday afternoons with her, tucked up under her cashmere blanket in her sunny office with harbor views, I'd want to dissect the how and why of the way in which meditation worked. Am I doing it right? How do I know? Where does it leave me and my cerebral ways if I reject thought? But her response was always, infuriatingly, the same: Just meditate.

Frankly, I've not the patience to share with you the countless studies that show that meditation works. This is not that kind of book. Just know that it does. Or Google it.

I do, however, want to share this: I'm crap at meditation. But for the past seven years I've meditated in my crappy way, twice a day for twenty minutes. I rarely "go down" into the deep place that others speak of. My experience is mostly chaotic and noisy. But here's the thing: You can be crap at meditation and it still works. The mere intention to sit with yourself is an act of self-care as far as our brains are concerned, which, *voilà*, triggers the comfort system. And, you know what? Even knowing it's okay to be crap at meditation is comforting.

And don't forget—anxiety is a head thing. It's characterized by thoughts. So, so many thoughts. Meditation draws energy down from the head. It works to still the mind. It turns the volume down on the thoughts. Meditation almost defies description. It's a wordless pursuit, so any description of it is a bit redundant. After years of meditating, I've realized words and thoughts can only point to the experience. They are not the experience itself. Just as the finger pointing to the moon is not the moon. Which can hurt the head when you try to think about it.

So . . .

Just meditate.

25. I met my current meditation teacher Tim in the midst of my Mid-thirties Meltdown, after Sky had moved to Japan. I was in a bad way, sick, unemployed, lost. I'd stopped meditating. Actually I'd never really started. Or it had never stuck, even after my attempts with Sky and years of doing yoga and various workshops and retreats.

When we meet in his Berber-carpeted rooms in the particularly refined Sydney suburb of Woollahra, I feel compelled to tell him that I have a prejudice against him. "You look like the preppy guy who looked down on me in my public school uniform on the bus. I feel scruffy sitting here." He did. And I did. Feel scruffy.

Tim smiled. "Meditation will help with that."

I came across Tim via a guy I met on Bondi Beach at 5am one morning. He claimed to know me from a yoga class and approached to tell me I needed to meditate. "I can just tell," he said in a way that suggested he regarded himself as highly intuitive. The tears streaming down my face as I paced the promenade predawn after an entirely sleepless night might have also helped him draw such a conclusion. He said I should contact Tim.

Later that week Tim was mentioned on two other occasions. Three strikes and I have to act. I guess it's a three thing again.

I learned the Vedic style, related to transcendental meditation, which originated more than 5,000 years ago in India and moved to China 2,000 years later morphing into the Buddhist tradition. It's a technique with structured boundaries . . . but then it lets you loose. You can sit in a chair, or on the floor. You do it twice a day. And after a shower is best because meditating releases an oil on the forehead that apparently makes your skin look younger and it's best not to wash it off.

You recite a mantra, faintly, in your head, for twenty minutes. That's it. If your mind wanders, return to the mantra. Don't worry about your breathing. Or your posture. Or your chakras. Return to the mantra. When thoughts bubble up, that's cool. Actually, it's better than cool. Thoughts are little pockets of stress that your consciousness encounters as it descends into calm. When you "think" them, the pockets of stress are released. *Pop!* And you return to the mantra. It's seductively convincing to know my thoughts are all part of the process. I'm not fighting myself.

But more than that my crazy, crappy thinking is what gets me to the meditative state. The more I think, the more I must gently return, in my case, to my mantra. In yours it might be to your breath, or to quietness, or to a speck of light burning bright in your third eye. Really, it doesn't matter, for it's actually the *repeated gentle returning to a quietness* that counts. It's this sturdy vigilance, this steering toward stillness, that builds the relaxation response—or calm muscle—in your being. And slowly, slowly you notice this calmness playing out in real life. Not immediately, but with time.

People like me are drawn to this Vedic style. It doesn't feel rigid. There are no rules and all is forgiven. And yet there are boundaries, a structure, encased in a discernible philosophy that can be pulled apart and played with and understood. I think this is why it attracts so many of us who are unable to just sit. The Beatles did this Vedic style. David Lynch has been practicing it for thirty-eight years and reckons his weirdest creations have emerged from his meditations.

When Russell Brand cleaned up his act a few years back (he started talking a lot of sense about the environment and ethical issues and I once saw him photographed in the gossip pages *riding a bike*), he put it down to Vedic meditation.

After a few weeks meditating with Tim I arrived at his studio in tears. "It's not working," I tell him.

"Keep meditating," he says. He pours me turmeric tea and sits back. "It's not really about what happens during the twenty minutes of meditation. It's what happens after, out there in real life."

"Right. This changes things. So meditation is like a little forum for airing grievances, purging the crap. So we can move on."

Tim doesn't say I'm wrong.

"You're watering the root so you can enjoy the fruit," he says and demonstrates a tree with his arm. "Keep watering, get the tree stable. And then things will grow from there."

When people ask me for the "one thing" that's helped with my anxiety, I tell them there's been no *one* thing. But if pressed, I concede that meditation has steered me to most of the good things that have happened in the past seven years.

The other thing I tell them is that the thing about meditation is that you always have it with you. You don't have to rely on anyone or anything. You sit. With yourself. And just meditate. This is incredibly powerful in itself.

HOW MEDITATION GOES FOR ME

I am not a meditation teacher and I don't want to share how to meditate here. Again, not that kind of book. I'm just offering a bit of insight into my experience. I figure it might help you feel more comfortable about it, and with being not particularly good at it. I still find it helpful to hear about other people's tussles with meditation.

I meditate after exercise and before breakfast in the morning. It helps when the body is "open" and alive.

I try to do it outdoors in the sun as much as possible. I meditate on rocks at the beach, on park benches in parks, on mountaintops at the end of a hike.

My head *always* meanders to my to-do list or to what I'll do right after meditation. In fact the whole meditation is a tug-of-war with an urge to schedule.

As this happens, repeatedly, I gently turn my attention away from the surging urge, to my mantra. It's like looking away from a kerfuffle going on outside to your right, away from the agitated conversation to your left, back to straight in front of you—no jerky moves, just a steady steering back to center.

My head wobbles wildly like one of those toy dogs on the back parcel shelf of a car. It only stops after I start to descend a little into stillness and the thoughts settle like those birds in the tree I described from that yoga retreat. If I do indeed descend.

Sometimes I'll open my eyes and I'm almost facing the back of the room. Like Chucky. I've bitten my lip before from a violent head jerk to the right. My teacher Tim, I know, watches me with a smile as I battle it out. Anxiety versus Me. Anxiety can sometimes still win.

Then there's this: The grimmer the environment, the better the meditation. I love meditating in cabs, in a parked car on a busy street between appointments, on planes during takeoff, in a sunny spot sitting in a gutter in an alleyway on the way to a meeting. During a stint working in TV, I'd meditate in the porta-potty while I waited for my curlers to set each morning. Working from a low base reduces the expectation. All that matters is that I'm sitting with myself. All that's left is the simple joy of, well, just me and just meditating.

This practice, repeated in meditation, now plays out in my life, I've noticed. I now gently turn away from kerfuffles and default back to "steady." My assistant Jo always comments on how I can shut off from calamities taking place in the office and window-washers and screeching fire sirens outside when we're having a meeting. I don't jump when there's a sudden noise. I can stay steady.

As the softness seeps in, even while thoughts do their jumpy dance in my head, it feels like the rigid boundaries of my body release. It's like I'm undoing a corset, or the button on tight jeans, and my insides are able to gently expand and my cells can stretch out languidly into the space created.

My teeth relax in their sockets.

The inside of my nostrils release. And if they don't at first, I focus on them doing so.

My fingernails soften in their nail beds.

My eyelashes soften.

I feel majestic and magnificent and suspended in a duvet-like cloud. Sometimes I get what I call my Michelin Man experience. I'm entirely convinced, my eyes shut, that my body has expanded several feet beyond myself in soft billowing folds, and I feel my "consciousness" expand to meet it. Everything that's rigid inside my body expands languidly into the softness.

If you own an Apple Mac laptop, you'll know that suction-y, shwooping thing the magnetic power cord does when it connects with the socket on the side. Well, when—and if—I finally arrive at the full, expanded, settled spot I describe above, that's the sensation. *Shwoop!* I fit. I'm connected.

When I come out of the meditation I try to hold this feeling. I open my eyes slowly and hold the gentleness. I stretch a little then stand up and keep holding. I try to hold it as long as I can—as I walk back home, as I have a shower, as I pack my bag to start my day. I hold it, I hold it.

STOP AND DROP

Not a meditator (yet)? Before I was able to get into it, Sky taught me a trick that was a good interim measure. "Stop. And. Drop.", she would say—by which she meant, stop your head and drop into your heart.

As I say, the thing about anxiety, it's all head. So anything that gets us out of our heads is good. It works a different muscle.

I used to keep a Post-it note affixed to my computer at the *Cosmo* office with "Stop. And. Drop" written on it. Several times a day I'd look at it and drop into my heart for a little moment.

You only have to hold the feeling for a few seconds to "get it." Try pausing your thinking for a minute and drawing your focus down into the space just behind your sternum. Do you feel the shift? Does a "knowing" ooze over you? You only have to touch it briefly for it to work.

26. **Sukshma [sook-shma]: 1.** *(adjective)* subtle (Sanskrit);
2. *(noun)* the practice of being innocent, faint and effortless.

In meditation you practice *sukshma* as you steer yourself back to the mantra over and over. Just say the word to yourself, innocently, faintly and effortlessly. It's so sukshma-ish, isn't it? Like when a child touches your arm when they come out at night to tell you they can't sleep.

You can't *try* to be innocent, faint and effortless. You just be it. Like, now. In this moment. I think this is the appeal of it for me. It's a rare thing in life that I *can't* plan for, or wait for. There's no run-up or tedious dress rehearsal. I just do it now as I type out this bit of my book (innocent, effortless key strokes) sitting in a public library with a heavy breather opposite me, having had two coffees when really I probably shouldn't have.

ROLL A SPONGE AROUND YOUR SKULL

If you're a regular meditator and anxiety makes it tricky at times, add this distracting trick to your usual mix: imagine a sponge gently working its way around the inside of your head, absorbing, mopping up the little anxious pockets. The mantra or breath moves the sponge around. You might find the inside of your head broadens.

DEEP BELLY BREATHING ALSO WORKS

Please note that meditation is *really really* hard when you're super anxious. It can be a bridge too far. The gearshift from a panic attack to a still mind is too dramatic. Know that this is cool. It truly is. So try some deep belly breathing instead at such times. A stack of science seems to support the practice. Dr. Richard Brown, an associate clinical professor of psychiatry at Columbia University and coauthor of *The Healing Power of the Breath*, says that deep, controlled breathing communicates to the body that everything is okay, which down regulates the stress response, slowing heart rate, diverting blood back to the brain and the digestive system and promoting feelings of calm.

Flick onto pages 155 and 156 for more techniques in such instances.

Deep breathing may also affect the immune system. Researchers at the Medical University of South Carolina divided a group of twenty healthy adults into two groups. One group was told to do two sets of ten-minute breathing exercises, while the other read a text of their choice for twenty minutes. The subjects' saliva was tested at various intervals during the exercise. The researchers found that the breathing group's saliva had significantly lower levels of three cytokines associated with inflammation and stress.

Harvard researcher Herbert Benson, who first coined the term "the relaxation response" in 1975 to demystify meditation, used scientific research to show that breathing counters the fight/flight response and can even change the expression of genes.

There are lots of ways people describe deep breathing, but I think the following is one of the simplest.

Sitting upright or lying down, place your hands on your belly.

Slowly breathe in, expanding your belly, to the count of five.

Pause.

Slowly breathe out to the count of six.

Repeat for 10–20 minutes a day.

If you have never practiced deep breathing before, you may need to work up to this practice slowly. Start by inhaling to a count of three and exhaling to a count of four, working your way up to six. Also, start by doing it for a couple of minutes and graduate to longer periods.

In some ways it works like meditation. The focus away from the head slows down new-brain activity. It also activates the comfort system. By voluntarily changing the rate, depth, and pattern of breathing, we can change the messages being sent to the brain. Also, by "massaging" our vagus nerve, which wraps our bellies, meandering its way around our organs and up to the brain, a variety of anti-stress enzymes and calming hormones such as acetylcholine, prolactin, vasopressin and oxytocin are released. Esther Sternberg, physician, researcher at the National Institute of Mental Health and author of *The Balance Within*, puts it well: "Think of a car throttling down the highway at 120 miles an hour. That's the stress response, and the Vagus nerve is the brake. . . . When you take slow, deep breaths, that is what is engaging the brake."

HAVE A GRATITUDE RITUAL (AS LAME AS IT SOUNDS)

I don't do vision boards, and I don't meditate to manifest handsome husbands and mansions with tennis courts. But I do have a gratitude ritual inspired by the time I shared a glass of tap water with personal development educator/self-help guru Dr. John Demartini, he of *The Secret* fame. I share it here because, like meditation, it's a daily practice that has had exponential impact on my anxiety. To be honest, it's a form of meditation.

Demartini is a central casting–issue Guru Dude. He wore black casual suiting (with a kerchief) when we met a few years back and speaks in dazzling motivational sound bites, generally trademarked. Indeed, Demartini owns The Gratitude Effect® trademark. He also has a cloth keyboard cover embroidered with "What You Think About and Thank About, You Bring About." I know both these things because, after discussing the topic of gratitude, he showed me the latter, along with his gratitude journal, a whopping great multi-volume tome that he carts around the world with him. He got teary as he read out a few passages for me. At the time I didn't know what to make of it, but I was forced to suspend my cynicism and try his gratitude ritual for the column I was writing. The results were wonderful. Really. And I practice it regularly.

My ritual is based on Demartini's, but a little less dazzling and sound-bitey. (And I've not trademarked it.) It goes like this: At night, after I climb into bed, I simply reflect for a few minutes on five things that pop into my mind that I'm grateful for. And say thank you for them. Usually they're banal things, like "Thank you for the flukiness that salmon was on special the very day I go to buy salmon!" Or "Thank you for my mate Rick who called today just to say he missed me." Who am I thanking? I guess it's the "universe." It might be God for you. It is for Demartini.

I don't seek a result. But it feels super good doing it. I asked Demartini why this might be so. The simple act of reflecting for a few minutes (he prescribes, pedantically, 4–15 minutes) on the good stuff in our lives creates a congruency between our goals and their fulfillment. This moment of recognition that things are gelling cooperatively makes you feel synchronicity and oneness with the flow of life. Which feels good, really good. You touch it, right? You know, the Something Else.

It's as if in that moment of gratefulness, everything makes sense. We realize all is okay and the world and the people in it are working perfectly, and we don't need to interfere for it to do so. For me, this is a massive, gulp-for-air feeling. The bigness of life whacks me in the solar plexus. Which is why many of us cry when we're grateful. I know I do.

Alex Korb writes in "The Grateful Brain," "Gratitude can have such a powerful impact on your life because it engages your brain in a virtuous cycle. Your brain only has so much power to focus its attention. It cannot easily focus on both positive and negative stimuli." Literally, you can't be grateful and anxious at the same time. Once again, the threat system in our amygdala is overridden.

On top of this, research shows gratitude stimulates the hypothalamus, a part of the brain that regulates anxiety.

Korb adds that the brain loves to fall for the confirmation bias—it looks for things that prove what it already believes to be true. "So once you start seeing things to be grateful for, your brain starts looking for more things to be grateful for."

And thusly we build all kinds of right muscles.

slow . . .

27. In my late teens I had a handwritten note Blu-Tacked to the back of my bedroom door: "Climb a tree." When anxiety struck, it would remind me to abort the downward plunge and scamper to a bush reserve fifteen minutes from the student house I lived in and get myself up a tree.

I'd sit up there, in my tree, sometimes for an hour, the after-work dog walkers passing beneath me, the eucalyptus sap causing a rash on my bare legs. I think it was the sheer ludicrousness of it all that lifted me out of the spiral.

I still climb trees sometimes, for the same kind of absurd jolt. But now I mostly go for a walk.

28. Before I learned to walk, however, I ran myself into the ground. That protracted Mid-thirties Meltdown I've mentioned a few times? It's probably time I shared how it went.

I'd been editing *Cosmo* for almost four years. For the majority of this time I'd been operating just outside my depth. I got the gig aged twenty-nine, a few weeks after moving to Sydney from Melbourne. I decided one day and within two weeks my little car (a Toyota Corolla back then) was packed and I was on the road.

I'd never read the magazine in my life, only glancing at it in doctor's offices. I'd never edited before either, nor managed staff. *Plus*, and, I soon learned, critically, I didn't own makeup or a hairdryer. Nor heels, nor eyelash extensions. I have to hand it to the then-publisher who employed me—she knew how to spot an overly earnest worker who would fake it until she made it and who'd just work and work until she *seemed* like all the other editors who'd attended the private ladies' schools and knew what Japanese hair straightening was. Because an overly earnest worker will just keep on working and working, long after she's faked it and made it.

I loved the job. The adrenaline rush sustained me and I was learning so fast. It was a sport to me. I got a kick from playing my position as best I could. But a few years into the gig things started to catch up with me. At the time, I was also extracting myself from a highly unhealthy relationship, the kind I'd warn young women about in the Love Life pages of the magazine. I'd shelved my friends. "We've lost you," my close friend Ragni told me one night over a dinner that had been canceled and rescheduled three times (by me). And I was drinking. A bottle of red wine a night. And smoking.

My fix was to go harder and to bang the proverbial square peg into the round hole with even more force. Because that's all I knew. I fixed things by going harder. And by running. I ran and I ran. I ran 6 miles each way to work most days. I ran to yoga class. I did sand-running races. I also competed in 24-hour mountain-bike races. By now I was sleeping three or four hours a night, going into work early and on weekends, trying to make it all . . . work. I was bone-crushingly tired, but I kept going.

There had been heart palpitations and I'd lost a lot of weight. My hair was falling out; I'd tug at a strand and more kept coming. The beauty editor brought in hair thickening powder for me to try.

And then . . .

We were doing a story in the magazine on the new "egg timer test," or anti-Müllerian hormone test, which had just arrived in Australia. The test claimed to enable women to see how many fertile years they had left.

Not one person on staff—and the team of fifteen were all women—was at the right point in their menstrual cycles to do the test for the "guinea pig" story we were running. Except me.

The test results came back a week later. The lovely lady on the phone was somber. "There's something not right," she said. "You have no female hormones left. None. It's like you're going through menopause."

My periods stopped the following month, after almost a year of irregularities. I was told I'd never have children.

The lovely lady advised me to see a doctor and endocrinologist immediately. I didn't. I shelved the prognosis along with everything else I couldn't face. And kept going and going, in the belief I *should* be able to get on top of things. The other stuff could wait. And, damn, no children? That was going to take a lifetime to digest. I wanted to leave that one on the shelf for as long as I could.

I hammered the square peg at *Cosmo* for another six months. Finally I quit, flimsily claiming "illness" in my resignation letter, although I'd still not seen a specialist. I was simply too fatigued to go on. I didn't share the exact details of my illness, because I didn't know them. But I've since been (helpfully!) told, you only had to take one look at me back then to know I wasn't well.

So did I see a doctor now that I'd pulled from the race? That would've been a self-caring and sensible thing to do. But I wasn't done yet. Goodness no! I still thought I could *work* myself out of my situation. Three weeks later I hopped on a plane to New York to write a book about a porn star. I'd been approached by a publisher to cowrite the book. Of course I said yes. It tends to be my default answer at the best of times. Yes! Yes! I lived with the porn star in her 42nd Street apartment and appeared in a hardcore porn movie playing a model that interrupts an NC-17-rated scene playing out on the casting couch. I believe it was called *Latin Adultery*, should you want to look it up.

Splendidly, the porn star got cold feet and pulled the pin on her memoir project, leaving me in considerable debt and without something to, oh you know, work on. No matter. I reached out to a bunch of magazines, and landed a commission for a story that saw me climb Machu Picchu for a travel story. It was enough to pay for my flight home. And enough to thoroughly exhaust me to the point of not being able to remember much of the high-altitude hike that many dream of doing. My only distinct memory is reading *MAD* magazine—the one copy I'd bought at a newstand in Los Angeles on the way through—with the photographer who accompanied me on the trip. Every night we'd lie in the dark in our tiny tent and laugh deliriously at the same gags. He'd fall asleep and I'd cry until dawn.

Eventually, eight months after getting the news about my fertility issues, I ran out of ways to flog the tired, sad horse that was me and went to see a doctor. I look back on all of this now and find the brazen signs I was being presented with ridiculously comical. I mean, you couldn't write a more clichéd script for a girl who needed to bloody well get real and stop climbing mountains. You couldn't inject more deus ex machina.

I went to get the results of my blood tests from the specialist. The doctor looked at his monitor then got up abruptly. He called out into the corridor and two interns followed him in. "Do you mind if they join us?" he asked me. I nodded; I didn't mind anything much at that point. I was numb with fatigue.

The doctor jabbed at his monitor. "I've never seen anything like it." The interns nodded and shook their heads. All three looked at me.

"Sarah, I'm not going to sugar-coat it. It's a miracle you're vertical." I didn't disagree. It took every last bit of energy to get to the clinic. Every joint hurt getting off the couch. I was filled with a gut-sinky dread when I tried to walk and I trembled from the spine outward just opening a door. I described it to others as feeling like three hangovers, a severe case of pre-menstrual antsyness and the kind of flu that sees you wince when the phone rings. All at once. When I tried to leave the house it was like in those dreams when you're trying to run from the swamp monster but you can't move your legs through the sludge, or where you're in an exam and you can't, as hard as you try, see the page. Actually, that sums it up perfectly.

"Well, I've been falling over a lot," I told the trio in front of me, and showed them the scabs on my knees.

The doctor told me I have Hashimoto's, an autoimmune disease of the thyroid. The thyroid is a small butterfly-shaped gland in the neck responsible for producing thyroid hormones. When your body cops a constant avalanche of stress hormones, your thyroid can get damaged and stop the thyroid hormone-producing party. Now, every cell in your body is affected by thyroid hormones, so when this happens, pretty much your entire body is affected, or at least all the parts of you that make you feel human. My doctor told me I have the worst case any of them had seen.

Further tests a few weeks later revealed my white blood cell count was barely existent, that I was prediabetic and that I'd developed osteoporosis in my right hip and in my neck. I was told, again, there would be no children.

"You left this far too long to get treated," the doctor said.

"What would've happened if I'd left it longer?"

"Heart failure . . . you're one or two weeks away from it."

(One endocrinologist told me I was "adrenally skeletal." "You're a Ferrari with a good duco, but you've been doing one-twenty in first gear. Your insides are metal rubbing on metal.)

"You need to stop," my doctor added and handed me a prescription for synthetic thyroid hormone, which I've taken ever since, albeit now at the minimum dosage.

29. Years ago at my cousin's wake, I stood eating asparagus sandwiches cut in squares on the veranda with the priest, and a mood came over me. It was a little waft. Everything went quiet. The curtains billowed. It was a particular mood. I'll tell you what it was. It was a big, vast, alive, connected, resigned, prickly, vulnerable *phew*. My cousin had been in mental anguish. He'd committed suicide at fifteen. It was all very harsh and sad, but I felt a softness come over the room as the fullness of life was felt by everyone there. Funerals, I find, often have that beautiful effect on us.

A similar mood came over me as I digested my diagnosis. I felt a soft relief. And, yes, that was it, I felt *excited*. What had just happened was a game changer. *This is serious, Mom.* This is another thing I say inside my head over and over (it's an OCD thing). For anyone born post 1990s, This is Serious Mom (TISM) was also the name of a band back before you were around. I say it when something big and important happens—like the three times I've fallen in love and the times when I've had big mountain-bike accidents and wound up in the hospital. I knew I'd been granted an opportunity here and that I had to rise to it, soft and full.

This is seriously big and awesome and overwhelming and I might not be a big enough vessel to fully process it all.

When the doctor told me the diagnosis, I knew I'd been given a wake-up call.

30. Here's why Hashimoto's is the perfect disease for people like me: it causes rapid weight gain (I put on 33 pounds in four weeks); your hair falls out in clumps; your nails peel off; you lose the outer third of your eyebrows (oddly); big angry pimples festoon your face; there's extreme sweating, stomach bloating, water retention, constipation, sluggishness and an inability to exercise (when you try to exercise, all of the above symptoms worsen); you develop debilitating indecisiveness (this has been linked to faulty thyroid function); this weird thing happens where foundation turns white on your face and makeup goes fluorescent pink; noises and bright lights hurt; and you just want to be left alone.

In essence it totally blows apart everything tight-fisted, adrenal A-types like me define ourselves by. It attacks our vanity, our pride, our emotional buttressing.

But Hashimoto's also serves a very important function. It stops us when we can't do it ourselves. It's like our bodies step in and say to us, "Well, if you won't stop, I will. And I'll collapse right here, in the middle of everything and prevent you from going any further down this path until you get a grip on yourself." When I share this in my public talks, the front few rows always nod their heads vigorously. That's another thing about autoimmune types—we're particularly earnest (what comes first, the disease or the behavior, I don't know) and tend to sit at the front, asking the questions, leaning forward, desperate to understand more.

When I was first diagnosed, it all hurt. Really hurt. And my anxiety flared (of course). I'd spent my life agile and I arrogantly traded on being fit and having a relatively androgynous form. I claimed to be "beyond my dress size"—a proponent of all shapes. But when I put on the 33 pounds I fairly and squarely regarded my body as repulsive. In some ways I think I always have. As a woman, I'm not unique in that. But being slim, not having flesh and curves and bumps to get in the way and impede dressing and moving, meant I could largely keep my body out of the picture. But now I had a disease that played itself out visibly on my body. This in itself dragged me down into an even darker place that I couldn't ignore.

Authorities tell us that 80–90 percent of autoimmune sufferers have anxiety/depression. The inflammation that the disease causes can lead to inflammation of the brain, leading to anxiety.

What causes autoimmune disease? There's generally a hereditary predisposition going on, which can be set off by a virus. Toxins, gluten, acidic foods, overexercising, the toxic effects of electromagnetic frequencies, fluoride and soy have also been implicated. The latest research is pointing squarely at inflammation and gut dysbiosis . . . and, *whattayaknow*, sugar as a truly toxic trigger. But for me, after years of looking into it and chatting to other sufferers, I think one of the most consistent root causes to be observed is the systematic hammering of oneself into the wrong holes. That is, running too hard at the wrong priorities, compromising ourselves, force-fitting ourselves into ways of living we know aren't right . . . all of which is highly abrasive and inflammatory. Some argue the inverse—that anxiety causes the inflammation, which then causes the various metabolic illnesses. Anxiety has been found to have the same DNA pathways as abdominal obesity, dyslipidemia, hypertension, diabetes and various autoimmune diseases. Either way, the correlation is there.

What fixes it? What's the only way to get back on track? The designated medications only do so much. They keep you ticking over. But to heal and thrive—both with autoimmune disease and anxiety—the only salve is slowing down, taking care of yourself, living cleanly and getting gentle and kind.

It's not coincidence that the approach is the same for both afflictions. For me, they are the same thing, both a symptom and trigger of each other.

31. Part of the deal with treating Hashimoto's was that I had to stop exercising. I could walk, the doctors said, but extremely slowly. For me, frankly, this entailed going back to the beginning and learning to walk again.

When you've only known two speeds—on and off—the mild mediocrity of walking pains your sense of self. I wanted to run. But even the idea that preceded the decision to burst into a jog *ugghhhs* at my head like a shrill laugh to the temple. Walking with the slightest hint of tension in my body or mind *ugghhhs*, too.

This became my barometer, this shrill but fuggy spasm of wrongness—a stinking mixture of regret, anger and disgust—that gripped me when I did anything that pushed or poked. If you've been really sick before, you might know this feeling. If you're anxious, you might know it, too. It happens, of course, to get us to stop, or at least slow down, and take care of ourselves. I know, I know. How bizarre, hey, that we would have such a mechanism! If you're someone who actually climbs into bed when they get the flu (that is, a life natural), you will, of course, find it very weird that the rest of us find this phenomenon extraordinary.

So I learned to walk and, along with meditating, it was a big part of my healing process, for both my Hashimoto's and my anxiety. It still is.

JUST WALK

When I'm anxious, or "thyroidy" (aka inflamed), I remember to walk. Not far. Not fast. Not fancy.

I breathe in for three steps—left, right, left—and out for four steps, like my deep belly breathing, but in motion. I count and breathe. I focus on drawing energy up from the earth, through my feet and up to the top of my head (two, three) and then I push the beige buzziness back down again through my legs, my shoes, into the earth (two, three, four). I breathe the energy in a loop, over and over. I move slowly and rhythmically like this.

After a bit it feels like I'm a conductor, fueled by the massive generator that is this rotating planet.

To do this you have to walk *reeeaaallllyy* slowly. Which is the point. Because all focus is shifted to the "breathing-and-staying-upright" part of your brain; the anxiety takes a back seat.

I remember one morning (after about two months of trying this technique) I was walking around the coastal headland not far from where I was living. It was early and the warm winter sun bounced off the ocean in sparkles back up onto my face and I realized I was smiling. I was smiling from my inside, effortlessly.

I didn't count my breathing anymore, but I still walked slowly and carefully, drawing energy up and back down into the earth. Blinded by a warm flash of sparkle I was aware I no longer felt heavy and ploddish. I wasn't springy either. I was kind of . . . just right. I didn't feel pulled down by gravity, nor did I feel like I was bursting away from it, countering it with calf-flexed bravado. I was kind of with it. In between it. And it was extraordinarily effortless. The *uggghhh* in my hips and in my head was gone. And I didn't want to get super excited about the just-rightness or start planning how I was going to build on this lovely, soft sensation . . . because that seemed slightly *uggghhh*-y. So I just walked.

32. A University of Toronto study looked at how this kind of mindful breathing while walking works to quash anxiety. Similar to compassionate therapy thinking, it found that we have two distinct brain networks. Our old brain sees us experience life as bodily sensations in real time. On the other hand, our new brain, which is our default network, tends to steer us to living life with a constant narrative playing in the background. Thus, we don't experience a warm zephyr or an awe-inspiring sunset in the now. Instead, the sunset triggers a narrative about how it's getting late and you really need to get home and defrost that damn pork. The study found that when you activate one network you dampen or disrupt the other. So, to bring things back to the beauty of the kind of anxiety-curing walking I advocate, it would seem that when you focus on the breath and the earth and the steps as a simple bodily sensation, you dampen the noisy, wandering storyline mechanism.

Another theory tells us that walking eases anxiety because it provides the surging stress hormones with an outlet. We were programmed to offload the build-up of stress hormones after the initial stressor was activated. A snake crosses our path. We freak. The hormones go into hyperdrive, telling our body to flee, fast. We flee and the hormones subside.

Lena Dunham, who's gone public several times about her anxiety, says: "To those struggling with anxiety, OCD, depression: I know it's mad annoying when people tell you to exercise, and it took me about sixteen medicated years to listen. I'm glad I did. It ain't about the ass, it's about the brain."

Studies show any movement, but particularly walking, will ease anxiety when we're in the middle of a stress hormone surge. Indeed, the studies show that a mere 20–30 minute walk, five times a week, will make people less anxious, as effectively as antidepressants. Even better, the effect is immediate—serotonin, dopamine and endorphins all increase as soon as you start moving. Please be aware that I'm not suggesting you flush your meds down the toilet and take up boot camp. I'm just highlighting the efficacy of exercise.

And while I'm adding caveats, I'll also advise against hardcore exercise if you're anxious. Gentle and slow stuff is best.

So . . .

Just walk.

33. I walk to work, to dinner with friends, to meetings. I schedule phone calls around my walking trips, to and from work. Sitting here I've just calculated I've owned a car for just five of the twenty-two years I've held a license. This is partly so I can walk some more.

There, another tip for having a better anxious journey: get rid of your car.

I've tailored my wardrobe to walking. Almost all of my shoes are flat with cushioned soles; my sartorial statement is Tracksuit Chic featuring shorts and stretchy pants and the kind of drapey stuff people like Gwyneth wear when they do "at home with" articles in magazines. I walk to black tie events in my fancy frock and flip-flops or lightweight sneakers that can fit in my oversized clutch after I change into heels around the corner. Oversized clutches are also part of my walking neurotic sartorial statement.

Blokes, you don't even face such impediments. So. Walk.

34. I'm also a mad hiker. I clear my weekends to hike. I head off on a train with a map and credit card or five bucks tucked down my bra and fling myself into the bush to traverse rocks and dust and creeks for a few hours. I travel to hike. It's my favorite thing to do. I set off over the Sierra Nevada, across Greek Islands, from English county to county. If it's a multiday affair, I set off with just a canvas shopping tote containing a change of undies, water, phone, toothbrush and perhaps a cucumber or two for snacking and rehydrating. Off I go for eight hours a day, a week at a time. Always solo.

Ingrid, my publisher, just added this comment: "My doctor calls it positive, neurotic behavior: you do it compulsively because you are neurotic but the net benefit is positive. In my case it's swimming." Positive neurotic behavior! I love that there's such a thing.

Friedrich Nietzsche and Charles Darwin did the same. They both hiked every single day, until old age. Both had anxiety. Both credit walking with taming their heads enough to be able to sort problems and bring their inspired ideas to fruition. "Only thoughts reached by walking have value," wrote Nietzsche.

Why does hiking work?

Hiking gets us into nature . . . and multiple studies show that folk who live in green spaces have lower rates of mental health issues. It's been suggested that getting away from city freneticness allows the prefrontal cortex to take a break. Accordingly, stress hormones, heart rate and other markers back off. Japanese scientists call the phenomenon *Shinrin-yoku*, or "forest bathing." Their recent studies suggest the benefits come from breathing in "aromas from the trees" known as phytoncides, an array of natural aerosols that trees give off for pest control.

One study found that salivary cortisol levels in people who gazed on forest scenery for twenty minutes were 13.4 percent lower than those who did the same in urban settings.

Hiking connects us to ourselves. A University of Michigan study found that because our senses evolved in nature, by getting back to it we connect more honestly with our sensory reactions. Which connects us with our true selves, and enhances a feeling of "oneness." Or perhaps we could say, a Something Else.

Neuroscientists at the Berkeley Social Interaction Laboratory have also found that awe-inspiring natural experiences release oxytocins—the hormones that makes us feel warm and fuzzy and connected with others. Ergo, that urge to interrupt a mega natural encounter—arriving at a waterfall or witnessing a spectacular sunrise—to take a photo or tweet the experience back to friends and family.

Hiking calms. According to a study published in the *Proceedings of the National Academy of Sciences*, a ninety-minute walk through a natural area led to lower levels of brooding and obsessive worry. Brain scans of the subjects found that there was decreased blood flow to the subgenual prefrontal cortex. Increased blood flow to this region of the brain is associated with bad moods. Everything from feeling sad about something, to worrying, to major depression seems to be tied to this brain region. Hiking deactivates it.

The study incorporated a control group who spent that ninety minutes walking through a city. They reported no such difference.

According to a 2010 report in the *Journal of Environmental Science and Technology*, even getting out into nature for five minutes at a stretch is enough to give your self-esteem a substantial upgrade. And know this: walking near water seemed to have the biggest effect.

Hiking creates space and clear thinking. One University of Minnesota study confirms that expansive environments—including rooms with high ceilings—inspire expansive thoughts. Another published in the *European Journal of Developmental Psychology* looked at why walking, specifically, gets us cerebrally fired up. Normally, multitasking results in weakened performance. But it found walking uses a separate "well" of attention to busy-mind thinking, allowing our "thinking well" all the space and resources it needs to do its thing. To reflect. And to calm.

Hiking gets us present. And this is key. Let me explain. When you hike over a long period in tough conditions (heat, rocks, steepness) you *must* enter the moment. You *must* focus on the here and now. This is because as soon as your mind wanders to thoughts of the finish line or of what you're going to eat for dinner or whether you have enough water for the distance or how tired you're feeling, you lose your mojo. Your heart sinks. Your head goes *ugghhh*. Instead, you must keep trudging and enjoying the trudging. The crunch of the rocks underfoot. The cicadas. The smell of the fig trees. Thing is, you *must* keep present, to keep going, to not feel like you're going to throw up. As soon as I start trying to calculate how much further it is to go, I feel a stab of sickness in the gut and I'm forced back to the present. And to joy. It's such a gift! Time passes, steered and corralled by the *ugghhh*-y pain of future thinking.

And one's anxiety abates.

35. Australian author David Malouf wrote in his exploration of The Happy Life in *The Quarterly Essay* that humans are happy within limits. He clarifies that he takes happiness to be something closer to contentment than happy-clappy elation (and thusly I was able to read through to the end of the essay). He argues that we're capable of being content even in the direst circumstances . . . so long as things are brought in to "human dimensions." Things can be super grim, but if the scope of what we must endure is narrowed, we'll cope . . . and even find meaning and purpose within these confines. Soldiers will experience satisfying camaraderie and true belonging in the very narrow confines of a POW camp, for instance. On a less extreme level, a canceled flight can see you have one of the most peaceful afternoons reading a book while waiting at the gate.

Malouf writes that a big part of contemporary unease comes from having so much of our life occurring at a speed that our bodies are not aligned with. We may be able to cross countries in mere hours, and catapult ourselves through space, but ultimately, he says, "we are still bone-heavy creatures tied to the gravitational pull of the Earth."

Yes, we are bone-heavy creatures. And there is a pace that was set out for us. And a pace for discerning thinking. But, boy, do we push the pace these days. And then we wonder why we feel so at odds with life. I get asked all the time why we're more anxious today. I think this phenomenon has a lot to do with it. I read Malouf's essay when I first set out to write this book and I realized that this was a crucial muscle to build—the muscle that gets us back in sync with our core referencing point (our bodies) for understanding the world, with discerning thought and the beat of our hearts.

Walking and hiking does this.

Cooking does, too. The witchy, grandmotherly type of cooking where you prepare things from scratch and you treat it as a hobby, not a chore to be rushed through. Fermenting vegetables gets me back in sync.

Yoga works. Hot and slow and controlled by breath, breaking down to our bone-heavy pace.

And then there's sex. Sex helps anxiety. I like this passage from D. H. Lawrence's *Women in Love*. Anxious Ursula and Birkin have had a fight. Then they have sex. Fully engorged sex. Then they go downstairs and Ursula serves the tea:

She was usually nervous and uncertain at performing these public duties, such as giving tea. But today she forgot, she was at her ease, entirely forgetting to have misgivings. The tea-pot poured beautifully from a proud slender spout, her eyes were warm with smiles. She had learned at last to be still and perfect.

Handwriting is great, too. I handwrite a lot. Sometimes just to calm my anxiety. It slows things down and forces me to connect with my thoughts, the discerning ones. Typing on a keyboard, by contrast, is too fast and it jangles our nerves as our fingers rush to keep up with our minds.

It turns out that David Malouf, too, writes his work longhand before typing things up. So does James Salter, whose quiet, still prose does great things for me. Stephen King wrote *Dreamcatcher* longhand. "It makes you think about each word as you write it," he told CBS News. "You see more ahead because you can't go as fast."

It's a source of much amusement for my staff that I handwrite everything first, including this book.

HANDWRITE ON A NAPKIN SITTING AT A BAR

In bars around the world I've sat and written out my loneliness and nervousness. On whatever comes to hand—scraps of paper, the back of a menu. In fact, the grimmer and more inappropriate, the better. It lowers the expectations. The point isn't what you produce, it's the writing out. And connecting with what you're thinking or feeling.

E. M. Forster said, "How do I know what I think until I see what I say?" (In fact, this very section was written out with a glass of red wine on the balcony of a restaurant down the road after a fretty day of getting absolutely nothing done because my anxiety had stepped in and taken up residence.)

It works like this (and BTW you don't need the wine, or even to do it in a public place).

Write what comes to mind. For me, mostly it's illegible

and in Pitman's shorthand (from my journalism days); I'm not going to read back over it. Gosh, no. At first, you can become agonizingly aware of your anxious impatience and how you'd really rather things moved faster. Your mind skips tracks, but try to focus on the ink scratching and bleeding on the paper and soften the grip on the pen. The focus and softening starts to take hold. And in a little while, see if the words start to look like what you're feeling at your core. For me, the words unfurl and my niggling, yearning feelings unfurl, away from the high-octane frazzle that goes on up in the front of my head, the place where words come from when I type on a computer. I allow the discerning, mindful thoughts to emerge at their own pace. These are the kinds of thoughts that put things into calm perspective. They're not yabbering, frenzied, defensive, nagging thoughts.

I find it takes about five minutes and the frazzle backs off. A quietness descends like warm treacle.

the something else (part 2)

36. So what exactly is the Something Else that we yearn for and that leaves us anxious when we don't have it? I put the question out to readers on my blog. I invited those who suffer anxiety to chime in if they felt comfortable. "What is it that you're missing that causes you to fret and get restless? What is it that triggers your nervous loneliness? What do you yearn?"

I didn't have to ask, *do* you yearn. I got 468 responses. No one writes back, "A new Lexus."

> *I yearn for a complete sense of self; I'm not sure it's something I can find or something I just have to wait for.*
>
> *I want to be authentic.*
>
> *I yearn to find the real me. I feel I am missing a connection with myself. But the thing is I want to find it while "life-ing." I want to have yearning and be in this life.*

Everything seems to be fractured, rather than unified as my gut tells me ought to be the case. This stems from a yearning for the world to make sense, to fit together.

I yearn for life direction and purpose. My dad's illness made me question what I REALLY want to be doing with my life as I could inherit the illness and I don't want to waste time.

I want to wake up. I feel like a zombie going through the motions of work and married life and the real me is dormant.

I want to know the real me, even if I have no idea what the real me is.

To know the connection to a bigger force. To know that the universe has got this one.

It burns at me every day to know that everything I'm doing makes sense.

When I read through the responses I noted that most of the responders seemed to be suggesting the same cruelly ironic thing:

cruel irony #4 – We yearn for something even if we don't know what it looks like or if it actually exists.

37. The Germans have a word for this.

Sehnsucht: *(noun)* An intense yearning for something far-off and indefinable.

Me, I yearn to sit with myself comfortably on that little bench in the sun. I yearn to know what this caper called life is all about and to know that I fit into a bigger scheme. That it all makes sense. I yearn for the expansive, magnificent spaciousness that comes from this calm connection and from not needing anything more. I want to wake up to this truth and to stop the pretending we all go on with—that shopping is identity-forming, that owning real estate matters, that beating our peers is a worthwhile pursuit.

This is at the heart of my anxiety. Actually, more specifically, the fact that I can't grasp it, connect with it, live it, is at the heart of it. It always has been. I hope you know what I mean.

38. As I wrote this chapter, a friend introduced me to a few books and essays by Irish poet and philosopher David Whyte. In his book *The Three Marriages*, Whyte says we need to navigate three vital relationships in life: one to others ("particularly and very personally, to one other living, breathing person"), another to work, and another to one's self "through an understanding of what it means to be themselves, discrete individuals alive and seemingly separate from everyone and everything else."

Whyte believes these relationships all involve vows made either consciously or unconsciously and that we should work on all three marriages, not as separate entities that have to be pitted against each other (in order to find that elusive "balance" between work, social life and "me time"), but as a "conversation" where all three are equally important.

But, he says, the toughest hook-up is with our selves. It's also the most critical, because without it the other two are but desperate, wobbly, outward-looking clamberings: "Neglecting this internal marriage, we can easily make ourselves a hostage to the externals of work and the demands of relationship. We find ourselves unable to move in these outer marriages because we have no inner foundation from which to step out with a firm persuasion. It is as if, absent a loving relationship with this inner representation of our self, we fling ourselves in all directions in our outer lives, looking for love in all the wrong places."

I committed to meditating regularly as I wrote this book. Even in meditation, when I arrive at that special quietness with myself, I can't stay there long. I surface very quickly, back to my thoughts, wanting to be distracted away from my internal communion. My mind jerks away from my little mate sitting next to me. I'm sitting here wondering why.

Is it because we're scared of meeting our selves?

Is it that we're scared what our selves show us?

Is it the sadness we encounter from realizing how long we've neglected our selves?

And want to know something funny? The day after writing the above—the *very* next day—I'm at a café (distracting myself from writing)—and, there, at the table next to me is David Whyte. He's sitting with a publicist who recognizes me and introduces us. He and I stayed in touch and met up for lunch on a subsequent visit. He orders a beer. We're both annoyed by the noise in the café.

I'm able to ask him why we fear coming in close. "If we crave to touch this Something Else, to know it, to be connected, why do we also flee from it, from our selves, into busy-ness and distraction and, well, all the things that make us anxious?"

Whyte's take is this: "Because there's a silence and aloneness that accompanies a strong relationship with yourself. In that silence we see the truth of our existence and the shortness of life. And this is painful.

"Also, when we come in close, we become larger . . . and this requires change. We become more visible, and thus more open to being touched by life, and thus more likely to be hurt."

I concede most of us do fear all of this. It does seem easier to just run from the Something Else.

But it also hits me in that moment that no matter which way you head, there's always anxiety. We have an original anxiety that stems from feeling we're missing something, that there's more to life, that we need to know where and how we connect with life. But to sit with our true selves causes another anxiety, a lonely, exposed anxiety.

Then, if we flee this sitting with ourselves, we encounter the anxiety of, well, knowing that we're fleeing ourselves and truth. It's a quandary; an anxious riddle, as Freud referred to it. I guess we have to ask ourselves, which anxiety is worse? Or perhaps the question is, if anxiety is unavoidable, which anxiety will produce the better life, the bigger life, the more meaningful life? The better journey?

Whyte also shares this: women are more anxious than men, or at least seemingly so, for evolutionary reasons. "Men spent a lot of time alone, following the one beast all day, which is a form of meditation. Women fed off the community. This stops the radical aloneness, the kind required to go in closer."

Perhaps. Though I did read the other day in *Scientific American* that the disparity in anxiety between men and women is beginning to be explained at a cellular level. Studies (albeit on rats) are finding that the most basic biological processes involved in the stress response differ markedly between males and females, such that females respond to stress much faster. It generally takes twenty-one days to increase anxiety behaviors in male mice but only six days in females. The researchers speculate that females evolved this way since a heightened state of alertness and awareness best served them to protect their young.

39. We weren't always so bewildered by human anxiety and existential yearning.

In Plato's *Protagoras*, twin brothers Prometheus and Epimetheus are charged with the rather large gig of Creation. They're told to set the world up so that every species has a quality or gift that keeps them safe, such that the entire kingdom can exist in balance. Birds get feathers so they fly from harm, deer are blessed with speed and cockroaches get cunning. All is created with sustainability and fairness in mind. But Epimetheus arrives at humans and realizes he's run out of qualities in his sack of gifts.

He has nothing to give man—no fur, no thick hide, no fangs, no great weight. Bugger. He turns to his brother Prometheus, the more insightful of the two. Prometheus suggests a makeshift solution. Humans will have to survive by being the inventors of their own nature. They'll have to improvise, inventing their own furs and manufacturing contraptions with speed. And they'll have to remain restless.

You see, while we're incomplete and restlessly aware of the very fact that something's missing, it will keep us forever striving forward (making fur substitutes and contraptions). And, thus, secure.

Later theologians banged on about the necessary angst of spiritual inquiry. They refer to it as the "divine discontent"; a hunger in the heart, a stirring to expand and grow and get closer to what counts—the Original Thing, The Yearning, the Something Else.

Lucretius the Epicurean said, "It is this discontent that has driven life steadily onward, out to the high seas."

Augustine of Hippo said, "Our heart is restless until it rests in thee." By *thee*, I take him to mean, well, that entity that sits on a bench with us in our heart space.

In 1937 Carl Jung wrote, "A psychoneurosis must be understood, ultimately, as the suffering of a soul that has not discovered its meaning." He observed among patients who were particularly resistant to normal treatments a deep anxiety that nothing was right. They eventually found stability, however, through one form of spirituality or another. "They came to themselves, they could accept themselves, and thus were reconciled to adverse circumstances and events . . . This is almost like what used to be expressed by saying: He has made his peace with God." Acceptance, rather than a cure, is the goal according to Jung. The release and energy that comes from such acceptance allows the individual to tap into "the meaning that quickens."

The meaning that quickens. This sings to me as an explanation of anxiety.

New York Times conservative political columnist David Brooks (get used to his name, I refer to his mindful opining a lot) wrote a dissection of Lady Gaga's particular brand of passion (and get used to my referring to passion and creativity; I believe they're part of the anxious experience).

This line sung out to me: "I suppose that people who live with passion start out with an especially intense desire to complete themselves. We are the only animals who are naturally unfinished. We have to bring ourselves to fulfillment, to integration and to coherence."

A bacon dress is an expression of this exploration. So is our anxiety, I reckon.

40. Before I read Lucretius, I drew a cartoon about an amoeba. I was probably about fourteen. This particular single-celled being was bored and one day decided to crawl from the prehistoric soupy detritus onto land. "What caused this brave little amoba [sic] to leave his comfortable life in the swampy waters to enter the harsh, open air?" I wrote under a drawing of a formless blob sniffing upward. "A yearning." I'd learned at school that the tail of the tadpole-like original being eventually evolved to become a human spine, and the olfactory bulb at its tip became the brain. Ergo, our sense of smell is so direct and original. And so very linked to anxiety. The area of the brain activated during anxiety sits right next to the olfactory bulb, and studies at the University of Wisconsin show that when we're stressed, stuff smells bad. Certain smells can also trigger anxiety. In an instant. I can certainly attest to this interplay. The smell of cut grass sees my stomach drop to my toes—it takes me straight back to swimming carnivals, veritable mud-wrestling pits for female bullying. If someone swims past me in a chlorinated pool, two lanes over, and they've bathed in Cussons Imperial Leather soap at some point in the past forty-eight hours, I'm rendered vomitous and wary. It's a smell associated with childhood trauma.

We yearned our way to becoming human. We yearned our way out of our mom's womb to oxygenated life. It's painful: we scream as we push forward into it.

41. I find it wonderful that science is such a rich companion in these thoughts. Crudely, quantum physics breaks matter down to its smallest parts in a quest to find who we are and where it all began. Advances in mathematics have seen scientists break atoms into smaller and smaller particles, until, well, it turns out they arrive at a point where material particles—stuff you can touch—disintegrate. The line between particles and energy becomes blurred and we might see everything as connected to everything else. To put it clumsily, we aren't particles as such, we're vast connected energy, or waves that clump together in such a way that it looks like matter. Which leaves many scientists with this same question: What is this connecting energy that we can't touch or see? And why was it created? And by whom or what? The Yearning for The Something Else, the original source of our anxiety, remains.

bipolar ?

42. Interestingly, if you head to the Black Dog Institute's website, its bipolar questionnaire resembles a quiz to determine spiritual awareness.

It asks things such as: Do you notice lots of coincidences occurring? Feel one with the world and nature? Believe that things possess a "special meaning?" Read special significance into things? Have quite mystical experiences?

Also interestingly, quantum physics triggered my first bipolar episode. (I do so love how these things tend to flow into each other, clumping together to form a life storyline.)

I was first diagnosed with manic depression, or bipolar, on April 8, 1996. I was twenty-three and living in Santa Cruz, California. I was on a scholarship to the University of California Santa Cruz to study German existentialism and took a graduate course called Philosophy of the Universe, something that could only have existed in Southern California in the mid-'90s. The first day of class our professor, pure math academic David Chalmers, announced that we'd be required to develop our own theory of time for our final papers. He was Australian via London, deadpan and dressed in a Grateful Dead T-shirt; an even scruffier Michael Hutchence.

Now, how's this for some more universal flow. As I was writing this bit of the book, my publisher sent me an article that linked into this idea of "the yearning" I'd been talking to her about over a martini, our preferred book-meeting accompaniment. The article was about the work of—hey ho!—David Chalmers. Turns out Chalmers went on to become a leading disrupter in consciousness research. The year he dumped his existential challenge on our little class in Santa Cruz, he also published his book *The Conscious Mind*, posing the conundrum of the "hard problem." He argued that there are many quandaries surrounding the human experience, but most are easy problems and, with time and increased intelligence, we'll no doubt solve them, much as we did the true surface of the earth.

But the "hard problem"—what makes us conscious, and what's the origin of life (the thing we yearn)—is possibly one we will never "solve" as our brains may never be capable of it. Actually, my memory of things was that Chalmers didn't so much declare the insolvability of consciousness as argue that *everything* is conscious. He then countered the ensuing outrage from fellow scientists by saying, "Well, how would we know that it wasn't?"

Anyway . . . it was in Chalmers' class that I first encountered the notion of universal consciousness. That the Something Else may actually be the oneness of everything. And it was his time-theory exercise that, perfectly, was my undoing.

I'd been in the States about four months, and I'd been off medication since I left Australia. I don't know anyone on antidepressants or anxiety meds who doesn't try, or think about trying, to come off their drugs. I've done it multiple times.

I continue to do it. There are side effects. Many. The bulk of the drugs used to treat anxiety affect sexual drive and ability to orgasm. Tolerance—hitting a point where the drugs no longer kick in like they used to—is also an issue. You can up your dosage, and up it again, but doing so can be so damn disheartening. It's like you're going backward. Also, for many of us it can feel unnatural to be on drugs for something so correlated with our character, our very selves. It can feel like we're cheating, no matter how many times we tell ourselves that anxiety sufferers taking meds is no different to diabetics taking insulin. The sensation for me at various junctures has been one of numbness. And of muting myself. My *real* self. Of course I draw parallels to the 1950s "silencing" of female outrage (or "hysteria") with Mother's Little Helpers (tranquilizers).

And, as I stated a few chapters back, there's always that niggling doubt about taking something that is still so poorly understood—no one really knows how or why they work, or indeed if they do. So we keep testing whether we can handle life without them. I write this knowing I really have to add that coming off medication should only ever be done with the support of a doctor who can keep an eye on you, and with whom you should discuss your rationale.

Prior to leaving to the States, I had found myself in a sturdy spot. A good shrink and several years of counseling under my belt, switching degrees from law to philosophy, a kind boyfriend, acing a political internship with the Labor party and scoring a scholarship had anchored me and left me feeling secure for the first time in ages. It felt safe phasing out the Zoloft that I was taking. Although, if I'm honest, the fact that Zoloft wasn't subsidized in the U.S. also played a part in the decision. And, if I'm even more honest, my psychiatrist was not entirely happy about the idea.

The first three months were a riot. Santa Cruz smelled of chai tea and burritos mixed with salt spray and I sat in cafés surrounded by Grateful Dead groupies and read *Ways of Seeing* by John Berger. I rode my mountain bike fast down the hill from campus into the swampy flats where the Mexicans lived. I ate nachos on cliff edges at sunset. I hung with old-timer surfers who introduced me to Buddhist thinkers and I lived with five lesbians and their ten cats. Two of them were lap dancers during the week and then went to S&M parties up in San Fran on the weekends. Sunday nights we'd hang out in the kitchen and they'd tell me their stories. I would apply stage makeup to their bruises and pink welts.

Things started to speed up. I walked late at night in the mist that rolled in from the Pacific, dragged by the heat of the desert.

I stopped sleeping. What did I do at night? I crawled up and down the floor of my tiny studio and pulled dust particles from the carpet, row by row. Plus, I showered—sometimes up to twenty or thirty times in a night. And turned off taps and shut doors. I had a circular sequence. Shower. Return to check taps. This would contaminate my hands. So I'd shower again. I'd do this in sequences of three. Then four. Then I'd have to check that the door to the bathroom I shared with the adjacent studio flat dweller was shut. Four trips to the door to shut it. Interspersed with four trips to the bathroom to wash my contaminated hands. And so on. You can see how this could continue all night. There was no endpoint.

I was addled with obsessive-compulsive disorder. More and more requisite routines slipped in. Four taps of the pillow. Four rows of carpet loops. Then I upped to sequences of five.

Months passed. I struggled to get to classes most days and it all got louder. The hum of the sunlight on the concrete streets, the smell of people's emotions around me. I could feel it all, loudly and brightly. I could smell everyone, I could *see* their sadness—or their stillness, or their emptiness—as they walked past me. I could feel the energy of a grain of washing powder at the Laundromat. Everything. What do you do with so much stimulation? Where do you put it? I couldn't process it all fast enough.

I chased the sensations—all of them—out as far as they took me, because I felt this was my duty. To grasp it all. That's what my mania was. A high-energy chase up the thought and sensation spectrum. I can keep up the pace for months sometimes and be thoroughly in command of the action, spinning the plates all at once. Like a conductor, aware of every pit member's need and next move. I'm in commune.

The pattern is now familiar to me. I'm the kite flyer. I let out more and more string. The wind whips my kite about and it's thrilling and I want to see how far I can go, how much string I can let out. I know any moment now, up so high, the wind might lift my kite violently and send it spiraling. Up this high, I have little control, but I let out more string because when that kite is at full throttle, with loads of slack, it's a thrilling thing to watch.

Then. *Thwack!*

Always, eventually, thwack.

At my most manic God is there. By God I mean the vast Everything I tuned into. I felt my most connected with Him/It as I flew up, up and away. I talk to God. We're in on this thing together. It has to be God because everything is so big, like looking through a magnifying glass, but from a million miles away. Everything is somehow to scale, yet magnified.

Four months into my Santa Cruz stint, when I dived headfirst down the steep staircase of the apartment building I later moved to, it was to test my connection to Him/It.

My one-room studio was on the top floor and I flung out and over the staircase, tumbling down on my stomach.

I didn't so much set out to create calamity as to use it to look for signs as to where I was meant to head. Likewise the time I wound up sleeping in a puddle on a college camping trip in the woods. I pitched my tent without a ground sheet and kind of just ended up submerged in water. I didn't move; I stayed there all night in a numb suspension. And I can only say it was to see where it would lead me.

Then when I set fire to my apartment a few days later it wasn't intentional and I wasn't mad. I was accidentally-on-purpose seeing what happened if I walked out of the house and left my acrylic socks on top of the gas heater. I was touching extremes to see if He was still there. Bring it on! Bring it on! I also got mugged on a bridge over the swampy marshlands late at night when I went for one of my ramblings.

I stopped seeing people once I got too fast. It hurt my brain to pull on the brakes enough to talk to people. I stopped going to school.

It was the time-theory paper. It was too big and it swirled faster and faster until I could not even write my name at the top of the sheet of paper. I was frozen, but buzzing, like static—most days unable to leave my little room. I'm not sure how long I froze. I've gone back through my diary to check and, going by the wildness of my handwriting and abstractness of the dire, free-form poetry I wrote, I'd say about three months.

It was David Chalmers who notified the university that I'd gone MIA, and a representative eventually came and got me when I hadn't shown up for six weeks.

"I'm not mad," I told the university psychiatrist assigned to me. "I'm the sanest person I know." But you can be sane, as in be perfectly cognizant of what's going on, and be going mad.

I wished I *wasn't* sane, I really did. When you're sane you have to witness the whole bloody unraveling with your eyes wide open.

While I was writing this book, and before we learned about his vile sexual misconduct, I read about U.S. comedian Louis C.K.'s struggles with depression and how he resented being a rational witness of his own descent:

"It never stopped getting worse. I remember thinking, *This is too much for me to handle*. I wanted to give up. I knew it was my *right* to. But then a few minutes would go by and I'd realize, I'm still here . . .

"There was no escape from it. And I'd be a little disappointed at not being truly suicidal. I *hated* being 'all right.' "

The cruel irony of high-functioning anxiety, yeah?

But in the university medical center that day I guess I ticked some boxes on the little form on the graduate psychiatrist's clipboard. I think he actually enjoyed connecting the dots so neatly, and he gave me a notepad and crayons so I could go home and draw my emotions. Seriously? Had he thought this through? I couldn't even write my name.

I'd left my first boyfriend George behind to move to California. Dear, dear George. He was the chef and owner of the café I'd worked at from the day I turned eighteen and we started dating only a few months before I took off to Santa Cruz, the trip having been planned before we got together. He was as simple and grounded as I was complex and flighty. He was ten years older and we both loved eating. The relationship was put on hold while I was overseas, but it was George who finally came and got me. We drove across the desert in an old Dodge van. I painted my nails lime green (I only know this from the photos) and I sobbed every dusk.

Back in Australia I was put on a cocktail of antiepileptic

medication and mood stabilizers and ferried from psychiatrist to psychiatrist to discuss my family of origin and do some more drawing of my emotions. I also saw Eugene, the ninety-two-year-old hypnotherapist who taught me about building muscles. Dear, dear Eugene. Some days I would just lie on the bathroom floor for the entire day, unable to move.

43. Am I still bipolar? I decided to stop asking this question a while back. Frankly, I'm not sure if my manic episodes were endocrinal in nature, triggered by the string of autoimmune diseases I've had over the years, starting with glandular fever when I was eleven. Or whether the bipolar stress caused the autoimmune dysfunction. (The latest research strongly implicates immune and inflammatory mechanisms in the development of bipolar.) Whatever the causal flow, I always end up with that same beige buzz that speeds up when I fail to sit with myself.

I still have distinct manic episodes. My kite goes whooping up and up. I'm thinking and talking fast. I feel synchronicities. I'll cry for days over the bigness of life. I get big ideas and I have urges—to put a creative bomb under situations, to poke a conversation with a wild notion, to plunge into physically dangerous or promiscuous pursuits.

But I now allow myself to fly for a bit—because my highs can be a lot of positive, creative and constructive fun. They're also inherently me. I have, and always will have, an insatiable need for connection. No amount of medication can eradicate this and I find myself making big compromises in other parts of my life to accommodate this drive (such as accepting having to be single for much of my life). I read somewhere that manic sex isn't really intercourse. It's discourse. An intense, expressive outlet for contact and communication. I get that and I apply the same lens to a number of my "behaviors" that others (and not just doctors) have previously put down to "illness."

Jay Griffiths asks if bipolar is really an illness (as does anyone who's got it). Illness is a condition that impairs normal functioning. "But in the foothills of mania normal functioning is enhanced," she writes. She points to Stephen Fry's documentary *The Secret Life of the Manic Depressive* where he asks all of the contributors on the program whether, given the chance, they'd press a button to rid themselves of their bipolar. All but one says no.

Australian actor Jessica Marais has grappled with bipolar. I connected with her via a mutual friend to chat to her about "it." I asked Jess if she'd press the button to rid herself of her bipolar. "No," she said categorically. "It's what's got me to 'know thyself.'

"Even during those times when I craved peace and calmness—'I just want peace and calmness!!!'—I've kind of known the journey would see me achieve this on my own . . . one day."

So the journey through anxiety and her bipolar itself is what matters? "Yes," she says.

Jess adds at the end of our chat that her going public with her bipolar a little while back properly "kickstarted" her own journey. "I've had to learn to love and embrace the fact that parts of my 'mania' or 'anxiety' mean that I am compelled at times to connect on a profound level with other human beings. And to emulate that connectedness for fourteen-hour work days for months at a time—which my work requires. I guess that, as I've learned to not operate on autopilot with it, notice it, and make better choices around managing it, it has become a friend."

A prominent London counselor told me something similar during one of the many conversations I've started in researching this book. She said, "If you're not anxious, you're not paying attention." I'm the same.

Granted, I also work at ensuring the string on my kite is not let out *too* far. In the main, I can now accept and temper things. I can go for the ride, but ensure I come back in close again, and in enough time. I've had to learn to. At such times I pull back from coffee, meditate (even if it's noisy and scatty) and I try to offload my enthusiasm on nature. I'll go for a mad run on a mountain track, or a scramble around an ocean cliff, or climb a tree, rather than call a friend and suggest we jump on a plane to Los Angeles tomorrow. As Jay Griffiths says, "The flaring energy of mania craves expenditure."

44. When I was four, before starting school, I'd watch that kids' show with the puppet Mr. Squiggle, the "man from the moon" who'd come visit each afternoon from one of his space walks. I'd sit with my brother Ben in front of the TV with an orange plastic cup of sultanas and peanuts and we'd wait for him to come on. I'd make "raisin burgers," squishing the dried fruit between two halves of a peanut, and nibble them slowly.

Strung up and jangly, Mr. Squiggle emerged from his tin spacecraft and would transform scribbles provided by viewers at home into funny pictures with his pencil nose.

There was also Miss Jane, who was a real-life human and forever patient and calm in the face of Mr. Squiggle's nervous antics. Mostly Miss Jane was there to gently pull Mr. Squiggle into line.

"Miss Jane, Miss Jane, hold my hand!" Midway through one of his drawings Mr. Squiggle would get too excited for this world and start to float off into space. "Spacewalk time, Miss Jane, spacewalk time," he'd say with the urgency of a little boy needing the toilet.

To my five-year-old mind, Miss Jane was a warm comforter that envelops you when you're home sick from school on a wet day. She'd never roll her eyes or get exasperated. She'd just gently reach up as Mr. Squiggle jangled out of shot, and grasp his little puppet ankle, pulling him back down to Earth.

"Sorry, Miss Jane. Thank you, Miss Jane," he'd mutter. "What would I do without you, Miss Jane."

It strikes me how much I would love to have a Miss Jane in my life. A good deal of my frenetic A-type female friends who are always running out the door with several handbags and multiple to-do lists have partnered with Miss Janes—rock-solid, unflappable men who call out from the couch, "I'll just be here when you get home." They complement each other wonderfully. The kite and the kite holder.

But when you've got a mood disorder it's often different. This is the hoary deal—when you have a mood disorder, few people are heavy enough and patient enough to anchor your ups and downs. And if you're high-functioning in your anxiety, there are not many men (or women) out there who will actually take the kite string off you in the first place. And I do wonder if it's grossly unfair to ever expect them to be able to. I've often expected this of my partners. The expectation was too high for both of us, with all of them.

If you're truly going to live fully and honestly you have to learn to be your own Miss Jane to your jumpy Mr. Squiggle. That's just the deal.

closer

45. We're all getting more anxious, not less. You're not imagining it. In Australia, anxiety-related problems have increased from 3.8 percent of the total population in 2011–12 to 11.2 percent in 2014–15.

The figures are similar for the rest of the world. A major University of Cambridge systematic review published in 2016 found anxiety has increased dramatically around the world—by about 3.8 to 25 percent. The largest increase was found among women, young people, and in Western countries. The report concluded that anxiety has emerged as a bigger problem than depression, with the U.S. scoring the highest number of people affected by anxiety—8 in 100. That's an estimated 3.4 percent of the population. The most comprehensive research on this phenomenon, conducted at the San Diego State University, found that anxiety has increased steadily over the past eighty years, but concludes "no one knows why."

Teen anxiety has become a particular focus for researchers. The annual American College Health Association study found 62 percent of undergraduates reported "overwhelming anxiety in 2016—up from 50 percent in 2011." A similar study at UCLA has reported a spike in students feeling "overwhelmed by all I had to do." In 1985, 18 percent said they did. By 2010, that number had increased to 29 percent. Last year, it surged to 41 percent. Perhaps tellingly, the number of hospital admissions for suicidal teens has doubled in the last ten years.

Here's my (possibly) contentious idea: It's because we're going in the wrong direction. We're grasping outward for satisfaction, sense of purpose, and for a solution to our unease. When we really need to be going inward, where the comfort lies. Wrong way! Go back!

46. *Every man rushes elsewhere into the future because no man has arrived at himself.*

— Michel de Montaigne

In his middle years, Renaissance man and self-confessed neurotic Michel de Montaigne set himself up in an isolated tower in the French countryside with the express aim of finding peace and writing life-changing essays. But sitting down to his writing desk each morning his head ran wild, lurching out for external salves to the anxiety that churned at his core. Sure, he was no longer frolicking around the world as he'd once done. But he still fled from himself daily. It drove him bonkers—he couldn't settle, he was flighty—and much of his writing was dedicated to this torturous reaching outward to distractions, and to the future, as he put it. All of which resembles my own attempts to find salvation in an army shed in a forest. And Thoreau's (in a log hut near Walden Pond), and Bill Gates's (in a cabin where he has "think weeks" on his own, away from humans and technology) and Elizabeth Gilbert's (in Italy, India and Bali).

But ironically—or perhaps perfectly—it was by sitting in, and writing about, the messiness of it all that Montaigne found his way to peace, as well as notoriety as a writer of life-changing essays. As he wised up to this idea, he shared through his writing that freedom from the restlessness in our beings could only be achieved by actively resisting the pull outward and into the future, and instead learning to "stay at home."

Home, in case Montaigne and I haven't spelled it out well enough yet, being ourselves. Our selves that sit on that little wooden bench, waiting for us to join them. Always.

It's right now as I'm going through an anxious spell that the fullness of this realization has *fully* occurred to me. That is, right now as I'm writing I'm viscerally absorbing the notion of coming in close, as opposed to merely processing someone else's words on the matter.

I've had to accept these anxious spells while writing this book. I mean, what could bring on anxiety more than a deadline on a book that requires doing the one thing guaranteed to trigger anxiety—sitting in, rather than fleeing the niggling in your guts?

It's some time not long before dawn and I'm crouched on the dining room floor naked, embracing the inappropriateness of my situation. I'm renting out a stranger's holiday home on Sydney's Northern Beaches. It's a pink wooden cottage in the trees—a steep climb up rocks from the ocean where I swim the length of the beach most mornings. The house is filled with another family's bunk beds and shell art and faux pineapples on the mantelpiece. I've been living temporarily, tentatively like this, in other people's spaces for five years now. I operate from one—sometimes two—suitcases of belongings. It somehow seems appropriate that I stay in this flighty space, with no "home" as such, while I write a book on the subject.

I've written a lot about anxiety over the years. I've answered lots of questions from strangers and friends of friends, and from myself. I now know that my anxiety doesn't have to be caused by anything particularly fear-inducing. At least not to the normal eye. After more than three decades of it coursing through my veins, anxiety is sometimes simply in my bones.

Yep, anxiety can just be in your bones. No reason required.

Indeed it is tonight. It's been there for a week or so. Nothing in particular has brought it on. It's just there. It might have been one coffee too many this week. It might be a hormonal surge. It might be the moon, the wind, Mercury doing that retrograde thing. Who knows? And does it really matter? Right now I just have to ride it out. And watch it dispassionately. I know better now than to fret about where it might have come from. Otherwise I'd be getting more anxious about being anxious, or about thinking I shouldn't be anxious. Sometimes I just am. Anxious. Full-stop.

Tonight, trying to sleep, parts of my head are relaxed and my body is crying out to sink and surrender. I've not slept for six nights straight. I've taken a Valium. *'Salright, 'salright.* You really have to do what you have to do sometimes. Sleep becomes too important. When you have anxiety, you do learn to give up on all the perfectly Instagrammable notions of how life should be done. You just have to attend to survival sometimes.

I'm dopey from the drugs, but there's a deeper part of my viscera that surges forward, urgently, relentlessly. The foot is clamped on the accelerator even if the driver is dozing at the wheel.

Flooding through my head, through the Valium haze, are thoughts about the emails that I need to send in the morning. I come up with an opening line for my next chapter. I map out my route to work tomorrow. I come up with an idea for a friend's business, and the logo. And I work out the significance of one of Adele's lyrics. These thoughts happen all at once in an explosion outward.

And it's now, as I chat to you, dear friends, from my squat position, that I *fully*—as in fulsomely—realize that this surging, grasping forward, "out there" and into the future is a really integral part of the anxious experience. It's there, always, being fueled by our anxiety.

It's difficult when you're in your anxiety to pause long enough to observe this. But tonight I try to stay put and watch it. And I see quite clearly that when I'm anxious all my rushing, competing, frenzied thoughts are either a) plans or b) contingencies for what could happen. It's like I'm running from the me that exists right now. This me, as I currently am, is not good enough. A good life and the "right" me and the answers I seek are ahead in the future . . . and I rush like buggery to get there. Ironically—oh, yes, once again, and cruel to be sure—the rushing, as well as the inevitable, vast and ungrounded unknown of what's ahead, makes me anxious. Which makes me rush ahead even more (to get away from the anxious, uncomfortable feeling I have). And so . . .

We rush to escape what makes us anxious, which makes us anxious, and so we rush some more. — **cruel irony #5**

And on and on we go. It's mad when you think about it. But sometimes it's just in our bones.

Sigh.

(The funny thing . . . as I watched this happen in myself in real time, the surge forward backed off. And I fell asleep shortly after.)

47. Thinkers from Darwin to Freud describe anxiety as a grasping forward to fixes that make us feel safer about the unknown ahead of us. I read somewhere that Charles Darwin observed that due to their inability to conceptualize the future, animals don't get anxious, at least not in the same way we do. Sure, frogs and ostriches experience fight-or-flight responses like us. But their trigger is plain and simple *fear* in that moment and this fear is proportionate to the tangible threat involved. (Plus, of course, they're not aware they're fearful or anxious as such, and so whatever fear or anxiety they have stops at the initial trigger.)

Human anxiety, on the other hand, stems from an existential awareness of what that fear means—ultimately our future annihilation. Indeed since Kierkegaard (in the 1840s), the existentialists have sought to explain this particular human anxiety as the dread we feel when we realize life is finite. When this strikes us, things can look different—we're aware of the lack of meaning to our lives (we breed, then we die, the end), the infinite choice and freedom we have (given we, as individuals, are not that important) and other vast factors that seem pointless in the face of our inevitable death. Fear is a primal physical response; anxiety is both this fear and an awareness of the fear and what it means.

Me, I can feel this future-grasping in my body. When I'm anxious, I often do my neck in. It's from straining forward with my head, rigid and forced. And I have accidents that injure my right leg. I plunge desperately forward with this dominant leg (in Eastern and esoteric traditions, the right side of the body is said to signify our forceful masculine side). I've subsequently broken my right ankle, torn my ligaments on my right ankle, split open my right knee four times, requiring stitching back the already strained flesh four times over, and I've developed arthritis in my right hip.

(A PARENTHETICAL NOTE ON DEPRESSION)

If anxiety surges forward, depression is a clinging to the past. Depression is being mired in regrets, remorse and obsessing over what should have been. When I was depressed briefly in my late teens I would repeat esprit de lescalier moments in my head. I replayed conversations over and over, grasping backward to avoid the niggling in my gut.

Now I repeat imaginary scenarios that are yet to happen.

Lao Tzu is wrongly attributed with an oft-cited insight (the idea actually originated from Junia Bretas, a Brazilian motivational speaker) that, translated and paraphrased, becomes:

If you are depressed you are living in the past.
If you are anxious you are living in the future.
If you are at peace you are living in the present.

Depressed or anxious, it's the unknown that we are most petrified of, so we grasp and cling to the certainty of what's already happened or to the false security of micromanaging in our heads what comes next. Or both.

To this extent I think anxiety and depression are different expressions of the same thing—a severe discomfort with what we can't grasp, what we can't know. In other words, the Something Else that I keep banging on about. Indeed, some researchers in this field, increasingly aware of the fundamental similarities between anxiety and depression, argue that both may be facets of a broader disorder. Other research has indicated that the same neurotransmitters play a role in causing both anxiety and depression.

Some of us have depressed anxiety. Others have anxious depression. Ninety percent of patients with anxiety have depression, while 85 percent of patients with depression have significant anxiety, with anxiety almost always the primary condition. For me, I'm mostly anxious. Depression kicks in as an exhausted response when my anxiety goes way too far. Some literature suggests depression is a natural coping mechanism deployed in such cases to stop us from self-combusting from anxiety that's out of control. A lot of anxious folk I've met agree with this take, intuitively.

The depression I experience is faded Catholic-school-uniform maroon in color, moldy in vibe, and feels like head fog from sleeping on an electric blanket or sitting in a 1970s office suite that oozes stale cigarette stench from the nylon carpet. It's always felt for me like a big heavy blanket hung over my head, muffling the buzz and holding back all my dreams and drive. I hate this feeling. Anxiety, for me, is more painful by a long shot, but I prefer the sharp pain to the muffling. It seems more productive. I get ego-boosting pats on the back for the things I produce when I'm in the early stages of anxiety, even if the toll on my spirit is so dire.

And. So. I tend to draw on every bit of adrenal reserve to ricochet myself out of such muffly beigeness, invariably back into high-octane anxiety.

Novelist Matt Haig writes in his memoir about his experience with suicide *Reasons to Stay Alive*:

> *Adding anxiety to depression is a bit like adding cocaine to alcohol. It presses fast-forward on the whole experience.*

Aurelio Costarella is a leading Australian fashion designer who I think is a true creative. I've worn his dresses for various TV award functions. They are vessels of meticulous, slightly oddball perfection. I felt I knew the kind of guy he was just from wearing his dress. Then, years later, as I wrote this book, I found out he suffers the same afflictions as myself. I read his posts on Facebook as he went through Transcranial Magnetic Stimulation (TMS) therapy, a form of electrical stimulation to the brain. I reached out to ask him how he was doing . . .

> *I liken my battle with depression and anxiety to being on a see-saw. If I manage to get some level of control over my depression, my anxiety bubbles to the surface.*

As part of writing this book I held forums at SANE and Black Dog to help me really poke into the issues. I asked the folk who attended for their thoughts on this point:

> *Depression and anxiety at the same time is being sucked into a hole, in the dark, but with all your nightmares chasing you, so you run around and around the bottom of the hole but never get away from anything.* – Lisa Jane
>
> *I have experienced both . . . sometimes anxiety can kick me out of depression. But then it's like a yo-yo experience and I have trouble finding peace in the middle.* – Mazarita
>
> *They're frenemies with me stuck in the middle.* – Annette
>
> *It's sort of like one side of your brain begging you not to get out of bed with chains, meanwhile the other part of you barks like a military sergeant for not getting out of bed.* – Joy
>
> *Anxiety and depression make me feel as though I'm stuck in tar and can't get out, even though my heart has so many dreams and aspirations.* – Samantha

ASK YOURSELF, "WHAT'S THE PROBLEM?"

There's a bit in Eckhart Tolle's *The Power of Now* that I've always loved. It's a little trick that can bring you back in from the anxious surge outward in a palm-to-face instant. It helped me get all that "be in the present" stuff that my anxiety had previously stopped me from even being able to conceptualize, let alone *feel.*

I went hunting for the exact passage just now so I could share it with you. You might like to try it, right now. Not in the future!

"Ask yourself what 'problem' you have right now, not next year, tomorrow, or five minutes from now. What is wrong with this moment?"

He asks you to try it right now with a problem. Try it with a bit of your particular brand of anxious buzz as you read this.

Feel into the problem *now*; not in sixty seconds, not in two seconds. Now! Your head might jump fifteen minutes ahead. No. Now. Is the problem still there?

Nope. It's gone, right?

As Tolle tells it, worries don't exist in the now. Worries about the future or the past don't exist either—they're just narratives we create in the present. Practice asking yourself "what's the problem" often. See if you don't start to feel the anxious cycle back away. See if those startled birds at sunset don't begin to settle, softly, gently, at dusk. See if this gentleness is where you want to be.

48. And what about this. When we're thrust into it, we anxious folk can often deal with the present really rather well. It's worth remembering this. As real, present-moment disasters occur, we invariably cope, and often better than others. The day after no sleep, I get on with things. At funerals, or when I've fallen off my bike, or the time I had to attend to my grandmother when she stopped breathing, or whenever a major work disaster plays out leaving my team in a panic, I'm a picture of calm. Dad used to call me "the tower of strength" in such moments. I also don't tend to have a lot of bog-standard fear (as opposed to anxiety). In fact, I relish real, present-moment fear and actively seek it out. At the expense of sounding like a humble-bragging wanker . . . I hitchhike, camp solo, fling myself down mountains on bikes, break up fights in the street, scare away snakes, scoop up spiders in glass jars and dispose of them outside for neighbors, surf breaks well above my ability, etc., etc. Just don't ask me to "just go to sleep if you're tired."

Real disasters are a cinch compared to the shit we make up in our heads.

Actually, they're a relief.

When the future does arrive, we're always okay. And I think my tendency to seek out risky experiences is about wanting to be reminded of this.

49. I think when you kind of get settled with the idea that anxiety happens when we go out beyond ourselves then you really start to feel miffed about the current way we deal with anxiety. In essence most modern medicine and therapy has worked to the notion that the "fix" is out there in the world somewhere. I'm here. The pills and experts are over there. And in between is a chasm of despair and lack of self-esteem that I have to wade through.

And it's not just the Western psychopharmacological model of diagnosis and treatment that creates this heart-sinky chasm. A lot of behavioral and psychoanalytical approaches, self-help books, spiritual gurus, motivational tapes and, yes, wellness blogs can do the same. Even if inadvertently. Lou, a spiritual guide and self-help practitioner I met in London a while back, reached out recently to tell me about her "brutal" seven-month breakdown that saw her move back in with her parents. She is one of the most aware and honest practitioners I've met and, indeed, in her email she shared that healing herself from what was ultimately diagnosed as PTSD entailed separating herself from the self-help and esoteric industries. She wrote, "I noticed the industry is another system that tells you something is wrong with you and is about someone else giving you a 'fix' e.g. healing/happiness/peace/enlightenment as an end goal." You can't believe how refreshing it was to read that.

When the pill or the guru or the "fresh start" doesn't fix the problem and the buzz (the cry of the missing Something Else) is still heard through the chemical fug and positive mantras, you go out even further to seek more answers—different pills, different experts.

Because we're told the answer's out there, somewhere.

Over the years, you find yourself clambering further and further out on a wild, fruitless goose chase. You grasp outward, chasing the next fix. Then the next, then the next. You're like a junkie. Or a shopper hoping the next throw cushion or handbag or novelty ice-cube tray will give you the cozy, feet-curled-up-under-you-on-a-lovely-soft-couch feeling you seek.

And all the while you're being told there's something wrong with you that has to be fixed. All the while you're dependent on others' ideas about what's wrong with you. It spirals us out to the perilous outer limbs, leaving us to sway around in the breeze of the latest self-help fad. A long way from home.

I'm wondering, can you see that this chase takes you further and further away from yourself? Can you see that it also takes the whole discussion away from the Something Else that you yearn for? Because you *do* kind of know that the peaceful, chilled *knowing* you're after is to be found much closer to the trunk of the metaphorical tree.

And can you then see that this all turns up the anxiety dial even louder?

50. In my lifetime, I've grasped outward to many things . . .

To sugar and coffee. A sugar or caffeine rush lunges me away from the peace I seek. It takes me up then it brings me crashing down. And so I grasp out for more. Sugar is uniquely effective at keeping us in a clutching, grasping anxious cycle. I've already explained how it makes us anxious in the first place. But it gets worse. We're actually *programmed* to hunt it down—to grasp and grasp for it. Why? Because it's such a marvelous and instant source of fat (which our ancestors 10,000 years ago found helpful). The stuff is also addictive and we have no "off/full switch" for it in our brains (as we do for all other food molecules), so we keep going back for more and more.

To alcohol. At one stage, in the lead-up to the Mid-thirties Meltdown, I'd drink a bottle of red wine a night. It would be red wine at night to wind down. Coffee and sugar to ricochet back up. Red wine at night to wind down. And so on and on.

To new cities. It's been a pattern. I get stuck and heavy with my career or a relationship or an apartment or the circling in my head and I up and leave. I'll make the call and be on the road within days, sometimes hours. And then I wear it as a badge of honor that nothing sticks, that I can live, yes, out of one suitcase for two years at a time, as, you might have noticed, I've mentioned several times already. That's pride right there. Ugly pride. It occurs to me now that external grasping, even if it's clearly dysfunctional, is mostly condoned as brave. Something to be ugly-proud about.

On the weekend just gone, my gorgeous friend Poh told me over dinner that I'm always fleeing. "You are sitting here, at the ready, about to flee to something else, away, onward," she said. "You need to get heavy." Blokes have said the same thing to me in regards to my apparent "inapproachability" from a dating perspective. One guy put it to me that I always seem to be on my way to somewhere else. "It's too intimidating to approach someone who might dart off on you," he said. I truly hate that I trigger this feeling in others.

So much about anxiety is about fleeing. We are given anxiety as a handy survival mechanism that jerks us into gear in the face of danger. But we abuse it, leaning on it when we're challenged.

To the Hare Krishnas. Oh, yes, this was one of my darker experiments. I lived for two months in an ashram in a dank forest desperately hoping the chanting and withdrawal would provide answers. I arrived in a dispiriting gray tracksuit that sagged in the bum and the knees and was told I was not allowed to think about sex (yeah, try not thinking of pink elephants). It rained torrentially. I ate mung. I turned to the skinny, orange-robed men who'd shunned the outside world and chanted for three hours a day for a way forward. I thought I had it all wrong. I thought the skinny men who wouldn't talk to me had it all sorted.

The sun came out one day and in the contrast this

presented I saw that the cocoon I was in was starting to smell like flannel pajamas that needed a wash.

I jumped in my car and fled. I don't recall saying goodbye. I went for a surf as the light faded. And then I bought some hot fries that steamed up the windows in the car.

To destructive partners. My second major relationship during my turbulent thirties, which left me unable to date again for seven years, saw me so far out on my metaphoric limb, deferring so much to him and his rather (okay, *very*) narcissistic ideas, that when we split my esteem was left to freefall for years. I ranted to a friend not long after our split, in earnest, "But he was the one in the couple who knew how life worked. What am I going to do now?"

To obsessions and compulsions. My OCD provides another grasping outlet. The counting, the checking . . . it keeps me doing stuff outside of myself, over and over. Ditto my bipolar behaviors. They're an unapologetic leap way out as far as I can go, often straight to the very thing I might be afraid of, like a big decision or a confrontational phone call or heights.

To the ping of an incoming email. When I'm anxious, I toggle on social media way past my 9pm blue-light curfew. I go into a zombie trance, numbly believing that the next like or find or tab will satiate me. I open up more tabs and create more options and overresearch and flag everything for follow-up. Collating, tagging, sorting, accumulating.

Plus, of course, to pills, doctors, gurus, self-help books and motivational wellness blogs. All of them.

I'm looking for something or waiting for something. But it never turns up.

51. I get Weekend Panic, although less and less these days. Weekend Panic is when you think you should be doing bigger things, farther out of town, all perfectly planned ahead. And the fact that it's Saturday morning and a whole heap of nothing is ahead of you sends you into a FOMO (Fear of Missing Out) spin. That's another thing I grasp out to—idealized downtime. I have this idea I should be doing the stuff folk in Country Road catalogs do (in crisp striped linen on moody beaches with fashionable friends and glasses of rosé). The fun is out there. The sense of purpose (which is important to anyone who finds it hard to have a sense of self beyond their working week identity) is out there. Everyone out there is having a more relaxing, rosé-colored time than me.

If I leave it too late to book the rustic cabin with the folksy throw rugs, or my fashionable friends aren't available, or if I'm too exhausted to organize the excursion, or if I finally get to the catalog-perfect weekend experience and the weather is bad and the cabin has rising damp and tubular furniture . . . I panic even more.

After years and years of this, fifty-two times in any given twelve months, I learned to back off with the weekend expectations. It took a long time. And writing several magazine columns on the subject to force myself to do it. If only as an "experiment." Now I venture to the local beach or pool. I do hikes that are a local train or ferry ride away. I leave Saturday nights for people who don't mind fighting crowds and stay in to watch Alfred Hitchcock movies and knit, perhaps ringing a brother or sister for a chat.

TRY A *FLÂNERIE*

Another such experiment . . . In Paris a little while back I noticed the locals don't walk around shops on a Sunday afternoon and buy stuff they don't need. Hyper-consumerism is deemed vulgar. Instead, they walk the streets merely to . . . wander and ponder. They call it a *flânerie*—a wandering walk. I once found a second-hand book with the cover ripped off called *The Flâneur: A Stroll Through the Paradoxes of Paris,* by Edmund White. To stroll in this way, White explained, is to be in real time with a city.

Parisians might drop in for a coffee or an aperitif at cafés where the chairs face outward such that they can watch other flâneurs. They visit gardens and poke their heads into avenues and parks and galleries. Just to absorb and look and reflect. It's a big part of the French psyche this simple observation of, and reflection upon, humanity. I love the spirit of it—sitting facing out to life. Then wandering among it. Then sitting back again. It's thoroughly absorbing, which allows calm, paced, discerning thought bubbles to surface.

When there's nowhere to go, nothing to do, we settle in close to ourselves.

When I get Weekend Panic, I'll walk to a bookshop, browse, walk home. I'll wander around a bit on the way home. I set my aims super low. My aim is simply to look at a few things, see what happens. You know, to enjoy staying close.

52. Let me quickly tell you about the time I ran into my mate Uge, a surfer I've known from around my neighborhood for a number of years. He was sitting in the sun having a coffee at a café. I asked what he was doing because he wasn't reading the paper or talking into a phone. He was just sitting. "Sez, I'm checking in with my Inside People," he said everyday-ishly.

I pressed him on this. He explained this entailed just sitting and asking of one's people, "Are we all happy? Comfortable? Heading in good directions?" We chatted about who these "people" were.

"Do you mean that side of ourselves we go home to after a bewildering day, or a loud night? The self we can see in our eyes looking in the mirror as we brush our teeth; the self that's always there, silent and knowing?" I asked.

"I guess so," he said.

Now for context, Uge used to have an office job. But seventeen years ago he started getting up at dawn to sit and watch people on Bondi Beach and take digital pictures of them surfing, running, meditating. He'd post them on this new contraption called a blog, which fellow desk slaves could access when they got to their cubicles in the morning. The message spread and soon enough he was able to toss in the day job and turned his site into a full-time gig. He now travels the world to surf and capture the joy of the sun coming up over beaches. Nice life if you can create it for yourself.

When I ask what anxious people get wrong, he's emphatic. "They don't give themselves time with their Inside People!"

HOW TO CHECK IN WITH YOUR INSIDE PEOPLE

I defer to Uge's advice wholeheartedly on this.

Don't overcomplicate things. Simply make the *time* to check in. Every day. Sunrise is his time. He honors it daily. Morning is my time, too. I carve out about twenty minutes to meditate. Or just hang out. Maybe on a bench in my head. Or I use a void in my day—while driving, while waiting at a bus stop—that I'd normally use to return calls.

It's pretty much meditation spelled out fresh. In fact, it reminds me of Sky's advice to just meditate. It's a powerful point. Just create the space with your Inside People and the rest will unfurl as it needs to.

Uge tells me that we then feel where our inside peeps are at. Try saying to yourself, as he does, "Are we good? Are we comfortable? Is this where we should be? Is it making sense?"

"Don't think or plan in this space, just check in," he says.

Then let stuff happen. It just does, "without trying," he tells me. In his case, a thriving business happened to him. Literally. It did for me, too, as I explained earlier.

Chatting to Uge I realized it's also important to *listen* to what your peeps tell you when you ask them how they are. It will probably be heard with a feeling, perhaps an expansiveness, a release. It's funny, for me, the answer that I hear is invariably, "Better than we thought, actually." Inside peeps are like that. When you check in on them.

53. What would it mean to stop? On an anxiety forum somewhere on the interweb a writer shares that his anxiety sees him "running on custard," a surface that is almost solid ground underneath him, but only if he keeps running.

"I must keep going . . . if I stop, the custard softens and I drown in custard . . . And the more people I talk to, the more I realize this feeling of exhaustion—of not being able to rest, of not getting anywhere—is common. And I wondered what it would mean to stop."

spirals

54. There's everyday beige buzzing or background anxiety, and there are full-blown anxiety attacks. I call my anxiety attacks "anxious spirals" because when they occur, they're not so much an attack, which suggests they're sudden and pounce upon me from out of nowhere. They're more a gradual downward, suck-holey momentum. My anxious spirals culminate when anxiety's beige buzz builds to a crescendo.

And you know, sometimes I actually prefer a dramatic spiral to the relentless torture of the buzz.

At least something is happening. At least the boil is being lanced.

A NOTE ON PHYSICAL "PANIC ATTACKS" VERSUS "ANXIETY SPIRALS"

I've become aware that many—nay, most—people's extreme anxiety manifests itself physically: it can feel like having a heart attack. A lot of the time when we hear about anxiety in the press, on websites, or anecdotally, it's this type of anxious blowout that is being referred to. Otherwise outwardly calm people will be walking along the street, or working out at the gym, or—and this seems to be the most common locale—meandering through a shopping mall, and then, *bam!*, they suddenly stop being able to breathe, their heart rate goes through the roof and they clutch their chest. They tremble, feel nauseous and dizzy. Such attacks understandably garner a fair bit of medical attention.

Personally, I've only really had this kind of attack once, when I was in law school in the middle of an exam. Since then I've instead had the pleasure of what I've been told are referred to as "intellectual anxiety attacks" (what I call anxiety spirals). These spirals are head-y. To the external observer I may look perfectly normal, but inside I'm a whirly-whirly of thoughts and nervousness. I'm not unaware of what's going on. Quite the opposite, I'm hyperaware.

I asked around to find out why there are two such distinct experiences. SANE Australia's Dr. Mark Cross explains that anxiety tends to play out on the body (somatically) when we haven't yet come to understand how and why our anxiety happens. This kind of panic attack happens when our thoughts trigger the ancient fight-or-flight mechanisms and we succumb to the response, believing something truly fearful is happening. In intellectual anxiety attacks (one of my spirals) we do the fight-or-flight response while simultaneously being able to understand what it's about. Not that this helps, because our overawareness of how and why anxiety happens and a thorough and genuine absorption in this feeds the spiral.

I chatted to someone at a dinner party recently about the distinction between the two experiences of extreme anxiety.

Anna, in her late thirties, had suffered a series of panic attacks following a bad breakup. She told me she had no idea what was going on and only learned later that it was anxiety playing out. I asked her if she's a nervous person normally. "No, not at all," she said. Which led me to wonder if the distinction is this: the chronically anxious have anxiety spirals, the otherwise laid-back have panic attacks during times of acute stress. Just an idea . . .

55. Either way, I'd like to share with you how my anxious spirals go. Actually, it's very much the last thing I've ever wanted to share and, until now, have avoided doing so. Only four people in my life have witnessed me going through one.

However, I reckon the telling of it nicely illustrates just how fundamentally useless external grasping is when you're anxious. I'm hammering home a point here.

My anxious spirals are mostly triggered by uncertainty and the lack of control such uncertainty entails. You might want to break yours down. Dig back the layers. What does it come back to? See if it ain't a fluttery, empty, unsupported belief that you just don't know what the hell to do, or what the hell is going on, or what the hell is *right*. For me, this will arise if I'm let down by someone or when someone or something leaves me hanging. It might kick off when, on a particularly wobbly day when the wind is not quite right and I've not slept, a significant other doesn't call when they said they would. Or it might arise from simply not being able to make a decision, like what to do with my Sunday.

Such triggers can leave me wobbling at the edge of the abyss of possibility where the limitlessness and uncertainty, if I don't nip them early enough, overwhelm me.

Yes, I know such examples are ludicrously innocuous. And, yes, it all does make me think of the starving kids in Africa. But that's yet another one of those cruel ironies with anxiety:

The more banal the supposed trigger, the guiltier and more self-indulgent and pathetic we feel, thus adding to the anxious spiral. — **cruel irony #6**

What do I do now? What's the *right* emotional or rational response? Where's the certainty? What do I grasp?

When it hits at night in the throes of insomnia, it's the perilous "Will I or won't I get to sleep?" factor that can send me spiraling. Will I be able to function tomorrow on no sleep? Will I have the energy to keep my team in the office confident in me? What will I do if it's 4am and I've still not drifted off? There's no certainty. No control. You can't force or micromanage sleep. Attempts to do so only push sleep further into the abyss.

Out there, wobbling at the edge of not-knowingness, I reach for external fixes that I can cling to, to stop myself from falling in, so I can get control of the situation and damn well sort this mess out.

As the uncertainty and limitlessness builds, I grasp to contingencies and intricate Plan Bs, Cs and Ds. "Okay, so if they don't call by 11am, then I'll call Friend X and Friend Y and see if they can meet me for breakfast instead. And I'll call Friend Z to ask what they think I should do." Then more thoughts come flooding in. But what if all this is a sign I'm meant to have breakfast on my own? What if this is my lesson right now and I'm ignoring it?

Soon I'm in a spiral of competing and complex strategies and fixes, freefalling. Some days I can slow things down, piece apart the thoughts and break the cycle. But on others the force is too much and down I go into the abyss.

I often ask for advice in these situations. More and more of it. When I met *New York Times* bestseller and TED sensation Brené Brown (we'll get to this fun encounter later), she said to me, "Asking for advice is a red flag." She works to green versus red flags. A red flag tells her that she's heading in the wrong direction, that she's in the wrong mindset and needs to stop and get a grip. I work to black and white versus color. If something appears in my mind's eye in black and white, it signals I'm being too rigid. I think anxious people tend to do this because when we're in anxiety it's very hard to access our intuition. For years I'd be told to "trust your gut" and "go with what you feel." But when you're in anxiety—particularly in an anxious spiral—you're all head. Blood rushes from your internals, powering the thoughts, disconnecting you from your gut.

In a hotel room, 30,000 words into writing the first draft of this book, I had an anxious spiral. You know, to keep me in the real-time, authentic grit and grime of it all. The Life Natural guy (the Tinder date who fishes) and I were in Hawaii. We were both traveling in different parts of the world and decided to meet up for an adventure. It was early in the relationship and we were going to have to share a bed for the first time. Until this moment, I'd been able to avoid this fairly fundamental step in the courtship process. It's all highly awkward, especially talking about it here. Despite years of work to destigmatize my fear of not falling asleep, I still rigidly control my sleeping arrangements with a white-knuckled grip. I feel I have to, to ensure I sleep, to ensure my autoimmune disease doesn't flare, to ensure I can function and run a business and write books and handle other humans and be a passable girlfriend. I wear earplugs, an eye mask and even tape my lips shut with surgical tape (a very cheap tip from my dentist to stop teeth-grinding). In hotels, I check for rattling windows and humming fuse boxes and bar fridges. I turn them off at the power point at night. Ditto LED alarm clocks next to the bed. Checking out the next day entails a detailed reconnection job.

I have to shower before bed every single night. This is entirely non-negotiable. Like, 100 percent. Since as long as I remember (at least since I was a small child) I've not gone without a shower before bed, not once. When I camp I have to carry in extra bathing water. My friends and brothers know this and factor it in to the backpack allocating. I've previously bathed in rivers with ice floating in them, in the snow in the Andes, in a dank, slime-filled swimming pool in Mexico.

On a hiking trip in Kakadu once, I bathed in a gorge while my friends Ragni and Kerry shone their head torches on the freshwater crocodiles on the opposite bank, watching their little red eyes for any movement.

The mere presence of another human lying next to me, their heart beating, their pheromones emanating, will keep me on high alert for seven hours. All of which is tricky to explain to someone early in a relationship. I'm ashamed to say I tend to work very hard to avoid having to address it.

But here we were, in Hawaii, late at night, with one double bed and two of us, and nowhere to dart to at midnight with a kiss to the cheek. And everything was up in the air.

My head desperately tried to solve the situation as a storm raged outside, rattling the plantation shutters. I juggled 387,462 possibilities in my head. I hadn't slept for several days. *What if I can't be a nice girlfriend tomorrow? What if my inflammation flares and I have to retreat to a dark room for the day and ruin plans*? I rang downstairs to see if there was another room I could move into. It was $700. *Is a night's sleep worth $700? What's the right answer*? I tried to weigh pros and cons, back and forth. I grasped at the Life Natural. I pecked at him for answers, reassurances, watertight guarantees. This was all new to him. He balked at my first rejection of his "you'll be right, there's nothing to worry about" efforts. He was not certain enough about this spiral he was witnessing and left me to stay with a friend.

I grasped at my stomach, clawing. I often clutch at my stomach when I'm in a spiral, digging my fingernails in and scraping away. (I ran a knife across my middle once, hoping the graphic splitting of flesh would render things more real and stop the thoughts. It's all I want—the thoughts to stop. But it only served to add more thoughts to the situation. *What will I tell others about this scar down the track? Where does this fit into the polemic on self-harming women?*)

I continued to freefall.

This metaphor works when I've used it to explain things: I'm Wile E. Coyote who's chased Roadrunner over the cliff edge, and I'm frantically treading thin air, trying to grasp at something solid to hang on to. But there's nothing there. Just the abyss. And the more I grasp outward, the more frantic I get. And down I go.

I also grasp at out-clauses. Here in Hawaii I clock the balcony, four flights up, several times. It's an option. When I'm alone in an anxious spiral, it's easier to find a way to slow down the spiral. But when there's the added fuel of another person's needs and confused face and defensive pushback (or even just the beating of their heart), the panic often worsens. I guess my base self (which is what I've descended to) perceives them as a threat. And my fight-or-flight mechanism goes into deranged hyperdrive. I'm so stuck, there are no more options I can grasp on to, that the only out is to . . . get out. To flee.

To. Stop. The. Thoughts.

Nothing else matters. In such moments I will often run. I've run 6 miles down a mountain in the dark with no shoes and no bra during my first anxious spiral in the presence of my first boyfriend George. I ran through Florence at 2am another time, with no idea where I was heading. Where and why was simply not something I cared about in those desperate moments. I just had to get out.

Loved ones try to understand it through their unspiraling lens. They can easily conclude that such "episodes" are either an attack on them (when their efforts to calm you have failed) or a cry for attention (also an attack on them—they mustn't be giving you enough attention). But it's not. I promise it's not.

After Hawaii, I had several more episodes like this with the Life Natural. I tried to jump from a car on a highway. I tried to jump out of a window in an Airbnb in France during a trip that appeared outwardly so Instagram-perfect. We'd been arguing in such a way that there was no endpoint and we were bringing out the worst in each other. We were holding mirrors up to each other's fears, mostly of abandonment if we're to get all Marianne Williamson about it. We'd descended too far, wanting the other to come to the rescue. And I could no longer navigate it and get us back onto dry ground. There were too many thoughts. I had to stop the thoughts. I had to flee.

Each time, it left the Life Natural entirely bewildered and angry and demanding an explanation. The shame and regret I felt afterward, and today, is indescribable. In part because the only answer I had then, and now with you, is that I don't know. It's all a big, bloody "I don't know." I don't know myself in those moments. I don't know why I can't stop the spiral. I'm smart enough to know better. But. It's almost like a short circuit occurs. Something very primal switches into gear. Everything tells me I. Must. Stop. The. Thoughts. And only something dramatic and powerful will do it.

Matt Haig shares in *Reasons to Stay Alive* what goes on for the anxious when they attempt suicide. "They could not care less about the luxury of happiness. They just want to feel an absence of pain. To escape a mind on fire, where thoughts blaze . . . to be empty." The only way he could escape his burning thoughts was to stop living.

A MEDIUM-TO-LONG NOTE ON HOW OTHER PEOPLE ARE REALLY HARD WHEN WE'RE SUPER ANXIOUS

I think it's worth acknowledging the Paradox of Other People (POOP?). Other people present another bundle of needs and thoughts and considerations to add to the goat rodeo going on in our heads. And they can get in the way of our control-freaky attempts to contain our anxiety when it starts to spiral, causing all kinds of cruel ironies for all involved. It's worth mentioning that, back in Hawaii, shortly after the Life Natural left, I splashed my face with water, pulled on jeans and went downstairs to the piano bar. I sat with a glass of red wine, wrote things out on a paper napkin, and chatted about love to the bar manager. Without the intense pressure of, and responsibility for, "the other," without someone else's confusion screaming at me when I just need someone to sit with me calmly, I could come in close and do what I know works to cope. All of which made no sense to the Life Natural when I explained later.

Here's a few more POOPs:

cruel irony #7 — The anxious tend to seek solitude, yet we simultaneously crave connection.

When I'm anxious every part of me wants to extract myself from other humans. I don't show up to things. I move to remote areas, away from everyone I know. I pack up and leave states, continents, relationships. I want to save them from the drama that is "me."

But the irony is, few things fuel my anxiety like being left alone with the buzz. If a friend cancels because she can't get a babysitter, I take this as social rejection. To me it's a sign that I'm a cosmic pain in the ass and that everyone is fed up with me and I don't fit and nothing makes sense. The very gist of why I jitter is the need to know I belong, that I fit.

We need easy-going people, but they can be our undoing. — **cruel irony #8**

We love easygoing folk. They can ride with our stuff. And they can be great teachers in the art of releasing a white-knuckled grip on life. But they can also tend to flake, and not realize what a big deal their flakiness is for someone for whom uncertainty can be their undoing. They can also defer too heavily to control-freaky anxious types. "I don't mind, you decide," they say. Which is lovely and easygoing, but also very challenging when you're organizing dinner for five such easygoing types and you're wobbly and, oh goodness, it all starts to tumble.

We cope with strangers better than our own mates when we're anxious. — **cruel irony #9**

I think this is because around loved ones we feel so bloody responsible and guilty and hyperaware of our inconsistencies and neurotic needs. It's exhausting being that apologetic.

In contrast, being polite and attentive with the old lady at the bus stop is like a job we must attend to. We busy ourselves with it. And this can distract us.

cruel irony #10 — We may come across as extroverted, but we have social anxiety.

I can stand on a stage talking to thousands of people. I can do live TV without having a conniption. Again, it's partly that I cope better with strangers. Plus, it's a job I have to attend to. I rise to the challenge. Like a chef I put on an apron, removing it once the shift is over. But if it's an everyday human experience that you're "meant" to enjoy, like a party, Lord help me.

I liked this from blogger Glennon Doyle Melton, who has to be physically alone to cope with her anxiety, even though she connects emotionally with her readers constantly:

> *Now, please understand that it is important for me to appreciate humanity and all those lovely humans who make up humanity from a comfortable distance. Because, close up, they all tend to make me quite nervous and often, annoyed . . . I am tired and socially anxious, so going to parties and showers and things such as this where I might actually be forced to sit next to and talk to humanity is really out of the question. So, I learn about love and humanity through books.*

cruel irony #11 — We can talk coherently and rationally about our anxiety, even joke about it, yet we freak out on a regular basis.

This is a cruel irony that affects our loved ones heavily. This explanation might explain the apparent contradiction. Anxious thoughts, apparently, have more pull in the brain than knowledge thoughts, so sensible facts and data go out the window when we're panicking.

We seem doggedly set in our ways, — **cruel irony #12**
but we have no idea what we want.

Our stubborn adherence to things (habits, rules, controlling triggers) is not based on a righteous sense that we are doing the right thing. Golly, no. We're flimsily coping, albeit with a white-knuckled grip. As I share in a later chapter, our anxiety leaves us totally unable to decide between competing preferences. If you're an anxious person's loved one, feel free to be firm telling your anxious mate what *you* want when they're in a spiral. They'll respect your preferences and respond well to the certainty. It's sweet relief. (And if you're an anxious person, accept this truth and go with a loved one's preference when they present it. That's the deal.)

We look strong and controlling. — **cruel irony #13**
But we actually need others' help more than most.

My control-freaky behavior creates the impression that I have everything sorted and, frankly, scares most people from wanting to approach me to offer assistance, even when I've gone AWOL or am standing in front of them, screaming out for help. The psychiatrist I was seeing recently, while writing this book, pointed out to me that I even micromanage how I receive help from loved ones. Which makes loved ones feel kind of redundant. And, yet, it's right at these precise moments I so desperately want someone to step in and convincingly take care of the planet for a bit. It's just that I can't correct my neurotic ways in time.

cruel irony #14 – We're always thinking about everyone (and everything), but we're so damn selfish.

I truly hate this about my anxiety. It can make me so terribly self-absorbed. I forget birthdays or don't have the energy or creativity to buy a present. And, yet, I wish I could explain that in my anxious moments I actually care more about the welfare of others than myself. Plane phobics are most concerned about their kids. Obsessive-compulsives are often scared that if, for example, they don't wash their hands, a loved one will die.

Oh, it's all just so hard for everyone, isn't it!

56. Another thing: I should say that sometimes the 387,462 competing, frantic thoughts tug at me from all directions until the net result inside my head is zero movement. I don't run or try to fight the thoughts with physical pain. Nope, I go numb. I suddenly go very still and stare into space and, if I stay still like this, there are, finally, no thoughts. It's now recognized that "freezing" like this is another common coping mechanism in the face of intense anxiety (psychologists now refer to the "fight, flight or freeze response"). We defer to it when there's no hope, like an animal that plays dead when it can't run any further, hoping its predator won't notice it.

I know other people who dissociate—they almost leave their bodies and, as one person I've met who experiences this way of coping describes it, see life through plate glass.

57. And another thing: When I'm in an anxious spiral I'm often keen to stay in it. I resist calming down. Stopping the runaway train seems too hard. And sometimes I wonder if I almost relish the drama of it, which is an idea that's truly irksome.

It's because doing the anxiety—and this is *really weird*—makes me feel safe. It keeps me in my frenetic doingness, which is just so damn familiar. I'm good at doing. Most anxious people are, right? Anxiety is a strobing party of doingness. I convince myself my doingness actually achieves *something*. When I'm anxious I'll often fan it by drinking more coffee than usual and sending out emails that complicate existing plans. I'm like those blokes who go out on Saturday nights looking for a fight . . . what, to have something to fight for? When in an anxious spiral, this reluctance to give in to calm can be heightened. If we stop our doingness, who's going to bloody well take care of this mess?

A smallish study on teens a few years back found that anxious brains were hyperconnected, which means that both sides of the brain "communicated" a bit too much. This led to overrumination, whereby a kid constantly thinks about a problem without actively attempting to find a solution, as though the rumination hits a pause on our having to dive into a commitment.

Dr. David Horgan, a psychiatrist and director of the Australian Suicide Prevention Foundation, says that the brain is basically a problem-solving machine—it looks for what is wrong, and then tries to think of ways to fix it. Unfortunately for anxious folk, the problems are not in the present moment—they're projected in the future—so we *can't* physically solve them.

Rumination, then, feels like we're doing *something*, at least. Anything is better than the nothingness of not knowing . . . and, I guess, ultimately, of having to sit quietly with ourselves. The doing, doing distracts us from the dread. It keeps us from sinking into the custard. Jodi Picoult says it pointedly: "Anxiety's like a rocking chair. It gives you something to do, but it doesn't get you very far." True, that.

Loved ones might try to pull me back from the brink, but if their attempts are in any way hesitant, or ill-conceived, or cautiously presented to me (which they often are, understandably), I'll resist. Their hesitation in the face of my frantic thrust outward only reinforces my need to stay in control. If I don't worry and fix and *do*, what then? Who will take care of things?

Besides—and I think this is key—in my funny little spiraling head, the anxiety ain't the problem, it's the solution. It's keeping this crazy ship afloat, people. Please don't take it away from me (us)!

58. So what *is* a poor loved one to do when faced with all this? When we're in primordial flight? When we aren't coping with other people? When we're being as complex as all get-up? When we wobble our way into a spiral or panic attack? Ohhhh, what a good, hard question.

Consider this the bit in the book that you flag and surreptitiously leave open for a loved one to read . . . but only once you fully acknowledge that your anxiety is not their problem. We have to be our own Miss Janes!

First, if I can extend this to anyone living in an anxious person's orbit, take charge when we're not good. I share this interaction with my mate Rick, who knew I was not good recently. And that I get Weekend Panic when I'm not good. It's perfect. It helped. And was not particularly onerous or demanding on him, I don't think.

> **Rick** (7:40am Saturday morning): *Hey Love. Fancy dinner out and maybe a movie? Are you happy for me to plan the whole thing and you turn up?*
>
> **Me:** *Yep!!! I'm going to use the above email as an example for how to make an exhausted control freak's day!*

Rick (11:40am, the earliest time possible to come back to me with confirmation, once cinema and restaurants open for bookings): *Ok movie and restaurant booked. Meet us at traffic lights at 6pm prompt!!!*

Another simple thing you can do, dear-loved-one-of-someone-with-anxiety, is to just be there, patiently, when we wobble. Just stay. And be entirely certain and solid about doing so, even in the very convincing face of pushback and the frantic wobbliness from us. Your patience and calmness will exist in such stark contrast to our funk that we'll start to feel silly and return to Earth. Our anxiety does pass.

TheMighty.com regularly posts things its anxious readers wished they could tell their friends and family when they're spiraling. The comments, from around the world, kind of echo this simple wish for you to simply . . . stay and be stable for us. Reading them might also see you feel less alone in your challenges with us.

Don't give up on me when I isolate myself. – Jen

Give me some space, but don't forget me. – Vickie

Get me to a quiet room where I can just be alone for a moment. My panic attacks normally happen because there's too much noise or too many people. So getting away is the best. – Amber

Help me to let time pass and let the panic attack run its course. Possibly, assist me in getting to a "safe" spot. – Kevin

During a panic attack, ask if it's OK if you come close. Getting in [my] face can make the attack worse. Sometimes holding my hand helps, sometimes it's a trigger. – Ashly

Keep yourself calm. I will eventually feed off your calmness and I'll be able to calm down. – Marissa

I need you to reach out to me, even when I'm so anxious I've stopped leaving the house. I need to know someone still cares and wants to see me. – Hayley

I understand you don't get it, but your efforts mean the world to me. – Avery

59. I, of course, ache for this kind of compassionate and resolved "I've got this one, babe" sturdiness from loved ones, even when (especially when!) I isolate myself, go numb, flee, run bra-less, push people away and all the other very ugly things that generally bewilder the handful of people who've seen me in a spiral.

But I do think it is too much to ask another human to fully understand the complexities of what happens when we go down. We anxious folk are fierce in our self-protection. We don't want our "fix" to be taken away. And we're very seductive in the art of pushing people away. I know I test others, to see if they can handle me. I think that's it. Or perhaps I'm just testing for sturdiness. *Please just make the decision! Please be the sturdy thing I can grasp as I spiral! Please just tell me in no uncertain terms that we're going for a walk around the block. And then we're going to cook nachos for dinner—definitely nachos because* you *feel like nachos (PLEASE don't ask me what I feel like!!!)!* All of which can be very confusing and demanding and testing for others.

And so I think we have to help them out. It's a responsibility.

My mate Lizzy can sometimes spiral into what she calls her "emotional cave." It used to freak her husband, Johnny, out and he'd run from the house, until she found a way to help out. "I told him when I go to the cave, it's not about him. I'm upset, I'm not upset with him. And then I said, 'I give you permission to come and get me from the cave.'" And this is the important bit that she added: "And I promise I will respond cooperatively when you do."

I reckon just the act of helping loved ones with our panic also sees the anxiety lift. It sees us draw on our inner strength, which we do have in bucketfuls by virtue of the fact we have to manage a bloody beige buzz all day. It sees us become our own Miss Jane.

60. The Life Natural rang me very early in our relationship. I was midspiral. It was a Saturday. I got all clusterfucky and confused and pushed him away. But he firmly told me he thought we should go fishing. "I want to go fishing, I think we should go fishing," he said. He gave me an address. "Can you get there in thirty? I'll be there with everything ready to go." I said yes. And my anxiety lifted even before I put down the phone. *Pfft.*

He was there at the appointed time with a beer and baseball caps. We ate bacon and egg rolls from the takeaway chicken shop. With a slice of cheese on top. Melted to perfection. So wrong and so right. "This is the best bacon and egg roll I've ever tasted," I sobbed. We caught two tiny fish that we threw back in.

My anxiety spiral lifted because a whole heap of firmness happened. A decision was made. There were sturdy details. He was sturdy in his desire to go fishing . . . he wasn't relying on me for a preference. Sturdy anything helps. It says to me that someone, something, has this one, babe. I feel held and this is enough for me to slow down the spiral myself.

But I'm going to be super honest here. The reason he was able to help me was because I took responsibility for helping him. To backtrack, I'd gone into a spiral because he was three hours late in ringing me to make plans for the day (hey, he's laid-back!). Which was just a trigger. But the uncertainty of not knowing what was happening and if I should call (and all the other uncertain early-relationship stuff that goes on) and the Weekend Panic and my anxiety about being anxious saw me go down, down, down. But I didn't say this at first. I just said I was in a panic. I was, simply, vulnerable. I didn't plant the cause on him. I just admitted I was hurting. And when he waded in and pulled me out, I made sure I responded cooperatively with a "yes." Later I was able to explain why I'd got worked up and to say that uncertainty and lateness and flakiness are triggers.

Sadly, it's rare that I've been able to hone in on what I need so well and steer things so responsibly. I try to remember this particularly daggy day in the beer often.

61. I followed a thread on an anxiety site one day that discussed the impact of us anxious folks' control-freakishness on partners. I emerged with this bit of advice to all the bewildered but caring partners out there: don't confuse our need to control our environment with a need to control you.

Another bit for loved ones to read. Again, you might like to flag it.

When we fuss and fret about getting wrinkles out of the bed, and ask you to double-check that you turned off the taps when you get up to go to the bathroom in the night, and ask you to stick to plans and call when you say you will, we're trying to control everything that we think might go wrong and that could trigger a spiral and ruin our time together. We're truly not aiming to control you.

And to all the partners out there, I get that it totally doesn't look this way to you. It's a massive stretch, I know. But I humbly invite you to perhaps *try* to see our intentions through this lens because maybe it will make you feel better about the very tough but noble situation you're no doubt in.

62. You know how I've shared the virtues of meditation? Well, I'll now admit it doesn't work in an anxiety spiral or panic. Our chaotic descents are fast, the adrenaline is spiked. Way too fast and spiky for anything requiring focused discipline and Zen-like centeredness. And to expect otherwise applies too much pressure. Once again, it's a bridge too far, leaving you wanting to stab someone's yogi cushion repeatedly with a fork.

Really, the only aim is to just come in a *bit* closer. In such frantic, spiraling moments, I find it's best to come in closer via the body. The body is solid enough, but not too "out there." It's close enough. I find my cells take over from there.

GET TOUCHED BY A SHOE ATTENDANT

I once had a wobbly moment on a Saturday in a pedestrian shopping mall. It was the Saturday-ness and mall-ness and general pedestrian-ness. I know that getting touched by a stranger works. It immediately brings my attention back to my body, which brings me in closer. It doesn't quite bring me to the wooden bench in my heart space where one can really peace out. But, frankly, when you're in a wobbly moment, meditation is often a leap too far. Baby steps, my friends.

I see a running shoe store. I dart in and ask the attendant to fit me for shoes. I tell him I don't know my shoe size. I do this for the express purpose of having him measure my feet with one of those metal foot-measuring devices. This of course entails him fussing awkwardly—but purposefully—with my feet with all the enthusiasm characteristic of every running nut I know. He does the talking, which is nice. He tells me about how he's studying engineering and runs with his girlfriend on Thursday evenings then eats pizza after. The touch brings me back in. I didn't buy shoes, but I did sign up to their runner's club, which pleased the running nut to no end.

PS: I think, too, that the off-beatness of doing something like this helps. No pressure, but don't hesitate either if you find yourself needing to step very slightly to the left to break a spiral. A little bit of crazy might freshen things up.

GET A THAI MASSAGE

According to a study published in the *International Journal of Neuroscience*, massage therapy decreased cortisol levels in the study participants by as much as 31 percent and increased serotonin and dopamine levels by the same amount. Scalp massages, says the study, are particularly beneficial. They send blood circulation to the brain and reduce the muscle tension in the back of the head and neck.

I also know this: Thai is best. It's firm. It's certain. It's low-commitment. With a Thai massage at a cheap joint, it's straight down to the business of bringing you in a touch closer. No tinkle music, no white robes and slippers. There's no pretense of it being special or fancy. Special and fancy just gets me anxious about feeling that I have to actually enjoy the experience.

I find the cheaper and grimmer the better. My regular is in the strip on Kings Cross, Sydney's red-light district, where I can hear junkies outside arguing over cigarette butts. There are threadbare Daffy Duck beach towels on the table and they play *Hooked on Classics* on repeat on the cassette player. Grim and lo-fi like this is a closer match to my predicament. It's not too huge a leap. I can, reluctantly, begrudgingly come closer.

63. Here are some more tricks as found on the internet, all of which are about coming in closer via the body so as to nip an anxiety spiral in its spirally track. Some are a little left of normal, refreshingly:

> *I take a big fluffy makeup brush and stroke my hand or my face with it. I keep a small size one in my purse for emergencies. It helps a lot if I'm out.*
>
> *The act of taking my hair down and then braiding it always soothes me. So does someone else braiding or brushing my hair.* [Oh, yes, me too!]
>
> *Wiggling! When I feel anxiety in my chest [and] it's really bad, I'll put on a song and literally dance it out. I pretend that I'm physically pulling the anxiety out of my chest, pull it or shake it out of my fingertips and slam it on the ground. This method has gotten me out of a lot of panic attacks.*

I rock back and forth in my rocking chair. Eventually I calm down and the rocking goes down to almost nothing.

One of the forums I did with Black Dog also revealed a few other techniques:

Counting steps helps me. Long strides are best! – "Book"

I read things forward and backward. – Gayle

When I'm feeling anxious and really not confident I just imagine that other people don't even care, which helps! – Joannah

I wear earplugs, to cocoon myself. – Kylie

A FINAL NOTE ON SOMATIC PANIC ATTACKS AND A GOOD WAY TO DEAL WITH THEM

If you tend to have the kind of panic attack that leaves you breathless, sweating and with heart palpitations, then see what you think of this. I found it really interesting and am able to apply it to my own experience of anxiety attacks. It's basically a process where you very clinically view your attack for what it is. It goes like this:

When the perception of a threat or danger travels to the amygdala, the fight-or-flight response is automatically triggered. This just happens. For us anxious folk the switch is particularly sensitive, of course.

But, you see, our poor old amygdala can't tell the difference between real and perceived threats. So, regardless of whether the threat is real (a tiger coming at us) or created by toggling thoughts or worry about tomorrow's team meeting, our bodies go through the same response.

Our pupils dilate to allow in more light and improve vision

so we can work out our exit strategy (judging how far away the threat is and the safest direction to run). Thus, we get blurred vision close up.

Breathing turns rapid and high in the chest, ostensibly to give the body oxygen to fight or flee. Thus we get tight in the chest.

But if we don't fight or flee, the oxygen builds up. Thus, we get dizzy.

Then our heart beats faster to shove the oxygen around the body. Thus, we feel like we might have heart failure.

Now we sweat to prevent the body overheating. Blood pressure increases. And our muscles tense.

And our veins constrict to divert more blood to the major muscle groups. Thus, our hands and feet can feel suddenly cold.

By now our digestive system has shut down so that nutrients and oxygen are diverted to the limbs and muscles. Thus, dry mouth, fluttery gut, nausea and diarrhea.

Okay. Pause.

Now take a moment to read back over those physical symptoms—the dizziness, shortness of breath, the tense muscles, etc. That's a panic attack right there.

But instead of viewing these symptoms as quite an understandable biological response that occurs in our brains, we anxious kids can tend to interpret them as evidence that there *is in fact* something dire and catastrophic and entirely threatening going on. In this light it could be said that panic attacks are a misinterpretation of symptoms. We mistake anxious-like symptoms for actual anxiety, which sees us get anxious about being anxious. Which can blow out into a separate syndrome called anxiety sensitivity, or AS, where sufferers become anxious about certain sensations associated with the experience of anxiety. It might be fear of vomiting or fear of shaking, or fear of having a panic attack in public. Oh, goodness!

I find it oddly comforting to know that sometimes my anxiety can be a case of misinterpreted physical reactions. It might be for you, too, especially if you tend to experience panic attacks like this, rather than anxious spirals. It dissipates the overwhelm, I think. And everything starts to back off a little.

Over time, I've come to find this phenomenon kind of cute, too. Like watching that red-faced kid at the sports carnival. We can have compassion and softness.

ASK YOURSELF, "WHAT'S THE PROBLEM?" (AGAIN)

I like this trick. It's a bit different than Eckhart Tolle's and possibly easier to deploy as you wobble on the edge of a spiral or panic.

The Happiness Project's Gretchen Rubin shared the trick with me over the phone during one of our chats. I think she refers to it on her blog or in one of her books, too.

She tells me about the time she got worked up about going to see her family in Kansas City for a week. She knew she'd have to work during the vacation and this made her kind of antsy. The tension mounted as the week approached. But on the plane there she did this:

She asked herself, "What's the problem?" She dug through a few layers. The problem wasn't that she had to work while on holidays. In fact she liked doing a bit of work before joining her family for the fun. The problem was that there was nowhere quiet for her to work. She dug deeper. At her parents' house there was no desk in her room. When she landed she went straight to Target and bought a $25 card table.

Planes are great places to force you down deep into taking a good hard look at yourself.

But, yes, the trick here, which I actively employ with a New York lawyer–type deliberateness quite regularly now, is to stop and ask, "What's the problem? Am I just experiencing the physical symptoms of anxiety? Am I?"

make the beast beautiful

64. Following the California episode in my early twenties it took about a year to build myself back up again. I couldn't study or hold down a full-time job. I worked on building mental muscle with Eugene, reading up on my illness (with real books, researched crudely, not via internet searches), and I waitressed. Waitressing is good for such occasions. It's bustling and distracting. You're in service, so a blissful eight hours can pass in which you don't think about yourself. And you can flee if you need to. You dump the coffee politely, then dash to the next order before your awkwardness freaks anyone out.

During this time, a guy I served coffee to gave me a book about obsessive-compulsive disorder. I'd asked about the red-raw rash on his hands and he'd told me he had a disorder that saw him wash himself over and over. I've never been one to hide stuff when prompted, particularly when I'm confided in; I presume I let him know I got his drift and he brought the book in the next day.

The book was called *Nine, Ten, Do it Again.* I remember stabbing at it with my finger, "Oh my God! Counting things is a thing! A thing that other people do!" So is doing it over and over. I recall reading that many OCD sufferers work to a counting rhythm of three, four and five. Electricity pioneer Nikola Tesla was a three man—before entering a building he would have to walk around the block three times and he would wash his hands three times. I, too, washed mine in sets of three. One, two, three. One two three. One two three. And then repeat, twenty-one times. Or ninety-six times. Or more. Unless I'd entered a four phase. I wonder now if it has something to do with the natural tempo of music, thus the seemingly lulling effect of counting for folk like me.

The very positing of this idea in my own head has just sent me down a horrible online rabbit hole looking into the various significances—mathematical, religious and otherwise—of the number three. For over an hour. I also learned, uncomfortably, that Tesla, like me, didn't ever settle and lived out his life in a Manhattan hotel room . . . on the 33rd floor.

Anyway.

The book exposed me to a few other factoids. Such as that OCD exists in the same numbers—about 1.2 percent of any given population—around the world, even in the depths of the Kalahari. The book also postulated (and I've picked up on this notion a number of times since) that far from being ostracized in ancient cultures, obsessive-compulsives were elevated to important leadership positions in communities. Their hyper-attendance to safety and hygiene—and all OCD symptoms cluster (in various, not always logical, guises) around these two themes—was a boon in days gone by. Shaman were likely OCD, goes some evolutionary theory.

I liked this.

About twenty years ago there was a documentary made about the work of Dian Fossey, who followed a tribe of chimps for several years. It gets cited in various guises around the interweb by people interested in the role of mental illness in society. The gist is that in all chimp troops, there always exists a small number that are anxious/depressed and that tend to retreat to the outskirts of the troop, often socially disengaged. Fossey decided to remove these agitated chimps to see what would happen. Six months later the entire community was dead. It was suggested that the anxious chimps were pivotal for survival. Outsiders, they were the ones who were sleeping in the trees on the edge, on the border, on the boundary of the community. Hypersensitive and vigilant, the smallest noise freaked them out and disturbed them so they were awake much of the night anyway. We label such symptoms anxiety, but back when we were in trees, they were the early warning system for the troop. They were the first to scream, "Look out! Look out!"

In *A First-Rate Madness,* Dr. Nassir Ghaemi argues that the best crisis leaders in history have had anxiety. "When our world is in tumult, mentally ill leaders function best," he writes. It's a bold claim, but he goes on: "In the storm of crisis, complete sanity can steer us astray, while some insanity brings us to port.

"The best crisis leaders are either mentally ill or mentally abnormal [he points to Winston Churchill, Abraham Lincoln, Martin Luther King and Gandhi]; the worst crisis leaders are mentally healthy." He says eminently sane men like Neville Chamberlain and George W. Bush made poor leaders. A lifetime without the cyclical torment of mood disorders, Ghaemi explains, left them ill-equipped to endure dire straits.

In the wake of the 2008 economic crash, some commentators have even suggested that the main cause was politicians and financiers who were either stupid or insufficiently anxious or both.

I absolutely believe it helps to see anxiety as having a metapurpose beyond the arbitrary torture of our little souls. Pain is lessened when there is a point to it. We know this. Women wouldn't go through childbirth and men wouldn't fight wars if this weren't true. For the anxious, this is possibly amplified by the fact that we tend to be very A-type, purpose-orientated kids who find the seemingly all-consuming, cruelly ironic, palpable pointlessness of anxiety unbearable.

During this same period in my early twenties, I also read Elizabeth Wurtzel's era-defining *Prozac Nation*. In it she wrote, of her depression, "That is all I want in life: for this pain to seem purposeful."

65. As Nietzsche said, "He who has a why can endure any how."

Our "why" today might just be the very important task of crying out, "Look out, look out . . . we're doing life wrong." We, the highly strung, are the advance party who flag to the troops that consumerism is hurting our hearts, that the toxins we're being fed via Big Pharma and Big Food are making us fat and sick and that . . . hang on guys! There's no triumphant finish line in this mad, frantic race. So perhaps we could, um, back off. It's we, the highly strung, who become meditation instructors, activists and online ranters.

New York Times bestseller and former addict Glennon Doyle Melton describes in a post how she was able to step out of the world of addiction by stepping "into worlds of *purpose*."

"Yes, I've got these conditions—anxiety, depression, addiction—and they almost killed me. *But they are also my superpowers*. I'm the canary in the mine and you need my sensitivity because I can smell toxins in the air that you can't smell, see trouble you don't see and sense danger you don't feel. My sensitivity could save us all. And so instead of letting me fall silent and die—why don't we work together to clear some of this poison from the air?"

I have often said the same—that we're proverbial canaries reporting back.

Glennon adds this: "Help us manage our fire, yes, but don't try to extinguish us."

STUDY SOME FRETTERS TO KNOW THYSELF

Can I ask you to do yourself a favor? Dig around. Google a bit. Find the point to your particular flavor of anxiety. Read up on Big Minds who contributed Big Things while anxious. There are many, and the correlation between creative contribution (artistic, political, entrepreneurial) and anxiety is well documented.

Poets, for instance, are up to thirty times more likely to suffer from bipolar disorder than non-poets. A 2012 study published in *Frontiers in Evolutionary Neuroscience* found that anxious folk tend to have higher IQs. Another study that year followed 1.2 million patients and their relatives and found that bipolar disorder is more common in individuals with artistic professions including dancers, photographers and authors. Scientists were also found to have the same link.

Simply studying anxiously Big Minds and Big Creatives can help us find our "why." Political philosopher John Stuart Mill, for instance, had a huge nervous breakdown that lasted eighteen months from the age of twenty. But he experienced a "small ray of light" when he started boning up on some French historians and reading poetry by Wordsworth. It got him more attuned to his emotions and more in touch with his inner emotional life. And this is what lifted his anxiety.

Consider the below a starting point for some know-thyself study:

Romanian philosopher and insomniac E. M. Cioran felt the greatness of humankind came down to those who didn't sleep. I read on the School of Life blog that he reckoned it was in those weird hours, out of sync with the rest of the world, that a singular creativity flourished, going as far as declaring, "What rich or strange idea was ever the work of a sleeper?"

Emily Dickinson's phobias left her housebound after the age of forty, which in turn left her quite a bit of undistracted time and space to write. Ditto Charles Darwin for several decades.

I follow Maria Popova's Brain Pickings blog. It alerted me to notorious fretter Anaïs Nin's diary notes about the importance of allowing her intensity to overflow as it needs to. "Something is always born of excess: great art was born of great terrors, great loneliness, great inhibitions, instabilities, and it always balances them."

– You should too!

Daniel Smith in *New York Times* bestseller *Monkey Mind* also experienced a discernible lift when he realized his anxiety was a necessary and important part of the artistic Jewish experience. "To be anxious wasn't shameful, it was a high calling. It was to be . . . more receptive to the true nature of things than everyone else. It was to be the person who saw with sharper eyes and felt with more active skin."

Frédéric Gros in *A Philosophy of Walking* offers this particularly grim insight from the mad hiker Nietzsche: "To those human beings who are of any concern to me I wish suffering, desolation, sickness, ill-treatment, indignities—I wish that they should not remain unfamiliar with profound self-contempt, the torture of self-mistrust, the wretchedness of the vanquished: I have no pity for them, because I wish them the only thing that can prove today whether one is worth anything or not—that one endures."

BONUS! A LITTLE LIST OF KNOW-THYSELF-BETTER READS, BY NO MEANS COMPLETE

The Road to Character — David Brooks

Your Voice in My Head — Emma Forrest

The Noonday Demon — Andrew Solomon

The Fry Chronicles — Stephen Fry

Monkey Mind — Daniel Smith

Reasons to Stay Alive — Matt Haig

My Age of Anxiety: Fear, Hope, Dread, and the Search for Peace of Mind — Scott Stossel

The Bell Jar — Sylvia Plath

An Unquiet Mind — Kay Redfield Jamison

Flick across the page — for a bit more about Jamison's story.

M Train — Patti Smith

Book of Longing — Leonard Cohen

66. Bipolar disorder, too, has a metapurpose, according to some evolutionary theorists. I've read many works by anthropologists who point out that manic depression is a genetic quirk that pops up in the same numbers across populations around the world—again about 1 percent—and might exist for an important reason. Daniel Nettle, in *Strong Imagination: Madness, Creativity, and Human Nature*, says manic depression is essential to the human genome. Those who experience intense moods are predisposed to building possible worlds, as well as to taking risks and testing boundaries. He explains that in the past, manic depressives pushed humans forward with their deep insight and creative urges; they strengthened the gene pool by who bravely venturing out of insular communities into uncharted territory. When they returned, they brought new skills that enhanced progress and survival. Nettle points out that these bipolar behaviors are respected if not revered in some cultures. In Inuit and Siberian cultures, for example, those deemed mad in our culture are heralded as healers and spiritual leaders.

Of course you can't get too haughty about such things. Declaring to those around me (who were quite convinced I was bipolar) that I didn't have a problem, but was instead one of a chosen few saviors of the human race, would have entirely disproved my point.

67. In my twenties, during my year of waitressing, I also read psychologist Kay Redfield Jamison's *An Unquiet Mind*, which details her own struggle with bipolar disorder.

A passage stood out for me:

"The Chinese believe that before you can conquer a beast you first must make it beautiful."

The Chinese proverb puts things in the imperative. I prefer to phrase it as a gentle invitation: Let's make our beast beautiful.

I believe with all my heart that just understanding the metapurpose of the anxious struggle helps to make it beautiful. Purposeful, creative, bold, rich, deep things are always beautiful.

In *An Unquiet Mind* Jamison comes around to thinking that acceptance, rather than transformation, is her endpoint, putting her ahead of the second-wave positive-psychology curve by a good fifteen years. I share this quote below because it's bold and true and has had a hugely significant impact on my own path:

> *I long ago abandoned the notion of a life without storms, or a world without dry and killing seasons. Life is too complicated, too constantly changing, to be anything but what it is. And I am, by nature, too mercurial to be anything but deeply wary of the grave unnaturalness involved in any attempt to exert too much control over essentially uncontrollable forces. There will always be propelling, disturbing elements, and they will be there until . . . the watch is taken from the wrist.*
>
> *It is, at the end of the day, the individual moments of restlessness, of bleakness, of strong persuasions and maddened enthusiasms, that inform one's life, change the nature and direction of one's work, and give final meaning and color to one's loves and friendships.*

Jamison takes things beyond a resigned acceptance of her unquiet mind. By accepting the storms and complications of her "individual moments" she's able to find a personal purpose to her life. Her beast becomes beautiful.

At the time I read this, I was grappling with my own bipolar diagnosis. But this passage said to me, "You know, it's okay." The storms and bleakness and madness count for something. The restlessness will lead to something. These "individual moments" or expressions count.

I've continued to explore this theme ever since and have found that my unique moments can either make me feel dreadfully alone and unhinged, or unique and a little bit special. It can be a choice to view your individual moments with bemused compassion and intrigue. To find them cute and beautiful. I try to do this. While trying to not lose connection with my humility.

SOME OF MY "INDIVIDUAL MOMENTS" . . . YOU KNOW, AS THOUGHT STARTERS

Do not in any way feel that you have to find these cute. I struggle to a lot of the time. But see if they can spark a bemused reflection on your own restless or bleak quirks.

I shower or bathe before bed. Always. But you know that already.

I change dinner venues on friends several times. If I'm not sure, it can become an ordeal that goes on for days as I try to work out the venue that will best suit all involved. I also change table positions, generally at least once. Mostly to avoid drafts or perfumed neighbors.

I have to go to the toilet three (or four) times before getting into bed. While counting in sets of three (or four). I know I mentioned that I'd worked quite heavily on my OCD back in my early twenties with Eugene. But I still live with a few stubborn vestiges of it that relate back to my fear of not sleeping. I accept this imperfect result. I call it a bit of residual scar tissue.

I overresearch. Everything. When I bought my most recent car, after being car-less for almost five years, I overresearched the most environmentally sound option on the planet. I couldn't help it. I looked into manufacturing carbon costs versus on-road carbon costs, how different driving techniques affected carbon loading and every other conceivable environmental factor. My overzealousness exposed a fault in the algorithms of Australia's green car guide. Which I promptly pointed out to the government department responsible. Who, equally promptly, returned my call to agree, yes, I was right and that they were correcting it immediately.

I took three years to buy my first couch. But rest assured it is, of course, the most functional, most toxin-free, most environmentally sound option on the planet. You might be interested to know Steve Jobs took eight years. His wife, Laurene, explained that they discussed the best design and philosophical principles of couches for close to a decade. "We spent a lot of time asking ourselves, 'What is the purpose of a sofa?'" Yeah, me too.

I have to do dangerous, reckless things occasionally. I call it "putting a bomb under the situation."

When someone loses something, I must find it for them. Every bin is searched, every beach scoured, until I do. Sometimes I'll picture its location (late one night, in the shower, or while riding home from dinner) two weeks later. *Bam!*

I can't settle for longer than about six months. My belongings have been in storage now for seven years. I could find this a problem. And regard my attachment to the notion as "ugly pride." But it's also who I am.

I still climb trees. At forty-three.

68. I did a presentation at a writer's festival not so long ago. It was a talk about turning a passion into a business. In the Q and A at the end, the first question from the audience came from a woman in her midforties who leaned forward over her notebook on her lap. She was half-standing in her eagerness to have her question heard. "On your personal blog you've written quite a bit about how you have anxiety. But I don't understand how you manage to do so much, to run a business and do public talks—and stand up there like you are now—when you have anxiety." She paused and sat back reluctantly. "How do you do it?"

I knew my answer.

I told her I look back now and can see that every major step forward in my career has been driven by my anxiety.

It leads me. It's my internal traffic light system that tells me "go" and "stop." When I feel the anxious choke at my throat, that *urghhhh,* I know something is not for me. *Stop!* it screams at me. In this way my relentless anxiety—and my awareness of it—has helped me make big important decisions along the way.

I think anxiety pushes us. It exists to do so—it helps us friggin' fire up. Even when it makes us stall with terror, it eventually makes conditions so unbearable that we ricochet off to a new important direction. Eventually.

This is what happened with my gig as cohost of the inaugural *MasterChef Australia* series in 2009. In this case, my anxiety didn't so much scream in my ear as explode. Throughout the eight months of filming my throat was tight with anxious choke. I was compromised by the role, shackled by the confines apportioned to women in Australian TV. I burned with frustration and boredom. It built up and built up and I'd stopped sleeping. The bags under my eyes, and weight gain from the sugary carbs I took to eating on set, saw me disintegrate in front of the record number of viewers who followed the first season. One day—the day we filmed the grand finale, aptly enough—I erupted. I was in the shipping-container dressing room I shared with my three male cohosts, being forced, again aptly, into a Jessica Rabbit corseted dress that saw my sugary carb–boosted bust billow voluminously. Did you see the finale? No doubt you watched it on wide-screen. My bust arrived on set. I entered shortly after. Ring bells?

The producers came in to the dressing room to announce yet another compromising set of instructions that would see me reduced to a vacuous talking head. I mouthed off at the producers for pushing us—myself and others who felt equally sleep-deprived and confined—too far. "I'm like a broom in a ridiculous dress with a wig. You press play and I mouth off your inane lines."

Then (I recall watching myself in slow motion) I punched the wall of the metal shipping container. I broke two knuckles. The big bosses were called in to have calm chats with me. I walked off set and went to see my meditation teacher, Tim, feeling deeply ashamed. I'd never behaved like this in my life. Tim laughed. "Perfect," he said. "You lanced the pimple, you were the volcano that released the pressure valve. Let's see what comes of it."

Sure enough, the next day, I felt resolved and clear. I quit, pulling out of future series. And moved to the shed in the forest and created my own business.

Anxiety is also the grist to my mill; the textured, all-weather athlete's track that provides the perfect surface from which I can make my highest jumps, up and over the rail. My anxiety activates my muscles, my fire, my fight. It also sees me care about everything. If I didn't care that the food industry was leading us all astray, if I didn't care that food wastage was killing the planet (and hadn't researched the bejesus out of such topics), I wouldn't have had the motivation to work seventeen hours a day to meet publishing and business deadlines.

If I didn't wake at 4am most mornings while building my business and writing my first few books, I wouldn't have got my product to the printers. Seriously. Those 4am awakenings, at times, have fine-tuned what I do. And if I didn't fret until things were the best they could be, my books wouldn't have sold. The public can "smell" when care is present, and when it's not. Social media is fabulous like that. There's nowhere to hide inauthenticity and followers simply drop away when they sense that posts are being done mechanically or by third parties or without passion.

Also, if I didn't know what it was like to go down deep and dark, as I have with anxiety, I wouldn't know how to take creative risks. The fact that I lost everything at several stages in my life to my anxiety and related illness means I have little attachment to material outcomes. I actually don't care about money. I don't get around to spending it. Quite frankly, having it makes me anxious. And so when I've come to big forks in the road in my career, I've been able to take the more unconventional, true path, unadulterated by bottom-line concerns. I've been able to rebuild and redefine my life several times because I've had nothing to lose.

And I should recognize that my manic outbursts, which I continue to have, have helped me write some of my most intimate, connected posts, often very late at night and in ecstatic tears. My mania, when tamed a little, has enabled me to connect, which has always been the basis of my business model. One of I Quit Sugar's five values mantras is to "Give a Shit."

I explain to the woman with the notebook at the writer's festival, that, sure, I don't know where to draw the line at times and when to rein things in. But that's my challenge: to use the anxious fire to ignite things, but then be able to dampen it to keep things burning steadily.

"You know that Lupe Fiasco song," I say, "where he kicks and pushes (and *coooasts*) and it's super smooth and heart lifting? That's what I do. I kick and push from the grist. Then I work on coasting. Which is much more fun than trying to fight anxiety."

69. Although, we mustn't overly romanticize things. *The Fault in our Stars* author John Green warned at a conference not so long ago against thinking we should all be inducing our anxiety (by coming off medication that otherwise tempers the condition) to access "genius." He pointed to the TV series *Homeland*—the main character Carrie, played by Claire Danes, comes off her meds so that she can discover the identity of the terrorists and save America. Apparently, she needed her mania to do this.

At the conference Green told the audience about his own similar experience the year before. He hadn't written a book in a long time and, blaming his medication, decided to go off it in the belief it would help him to write. The experiment didn't work: he produced nothing that made any sense during that whole time. In the end, he said, it was better to deal with his chronic problem, to manage it rather than romanticize it, in order to produce his best work.

Glennon Doyle Melton has a slightly different take. She goes on and off antidepressants because she needs the desperation of depression to fuel her creativity. On medication she says she loses the feeling "*if I don't write I will die.* This is how I feel when I'm depressed. Since I lose my joy and meaning, I come to the blank page to create meaning and joy, to get it back. Because I become desperate to make sense of things. And that desperation, I'm afraid, is what makes my writing good."

It's a fine line—at what point does respect for your heightened sensitivity and creativity become problematic—even dangerous? I can't really answer it. I, too, flirt with the agitated, creative edge. I've toppled, too. But I see it as a responsibility to keep alive to that sensitive tipping point at which I need to pull in (and take medication or get help or rest). It's a bit like a parent who must know when to rein in their hyperexcited toddler who's showing off to the guests almost embarrassingly (but creatively and expressively) before it ends in tears.

70. Just a small thought. I've learned that at a biological level, anxiety is a lot like excitement. Both anxiety and excitement make my heart quicken and my stomach flutter, and send a wave of "Ooh ooh ooh, This Is Serious Mom" over me.

So you know what I do now? I often *choose* to interpret anxiety as excitement whenever I can. Standing on the precipice, about to jump into something new, I often feel anxious. But if I pause and reflect, I realize it could equally be excitement that I'm feeling. When you see it as excitement it's BLOODY FUN.

Since handing in the first draft of this book, I've learned this thing that I do has a name—"anxiety reappraisal."

In 2013 Harvard University researchers found that simply saying "I'm excited" out loud could reappraise anxiety as excitement, which in turn improve performance during anxiety-inducing activities. Another study published last year asked people to list goals they had that were in conflict with one another, such as training for a marathon and finishing an ambitious project. Some participants were asked to recite the phrase "I am excited" out loud three times, while others just recited their names. Those who reappraised their anxiety as excitement felt they had more time on their hands to complete their goals.

Tiger Woods once declared, "The day I'm not nervous is the day I quit. To me, nerves are great. That means you care." Right on! Bill Russell, one of the best basketball players in history, anxious-vomited before 1,128 of his games. To his teammates it was a good sign—he was on fire. But then there are just as many sporting heroes—and war heroes and political leaders and artists—who can cite moments where their anxiety choked them. Greg Norman blew a massive lead with a big fat choke once and wound up crying in the arms of his competitor. Thomas Jefferson and Gandhi both suffered social phobia. Barbra Streisand choked for twenty-seven years, unable to perform live. Carly Simon had to stick pins into her skin before going on stage to distract her from her anxiety.

Adele has extreme stage fright and suffers panic attacks. She has developed an alter ego to cope with her anxiety and also limits her touring. She says she's scared of audiences and once used a fire exit to escape from fans. At another concert she projectile-vomited on an unfortunate audience member.

Actress Emma Stone says acting, funnily enough, is the thing that helps her. "There's something about the immediacy of acting," she said. "You can't afford to think about a million other things. You have to think about the task at hand. Acting forces me to sort of be like a Zen master: What is happening right in this moment?"

But again, it's a fine line between flying and choking from anxiety, between being a hero and a coward. I don't know how I've wound up using so many sporting analogies, but I have another one. Mike Tyson's trainer once told journalists, "The hero and the coward both feel the same thing, but the hero uses his fear . . . while the coward runs. It's the same thing, fear, but it's what you do with it that matters." Yes, it's what you do with it. And perhaps whether you treat it as beautiful.

When I'm doing something definitely scary, I'll allow my anxiety to have the stage. It's always got me through math exams and driving tests and TV gigs. I let it express itself. Which might sound irresponsible. But I've found that it's only when you put the brakes on its forceful charge through your system that it leads to things like freak-outs or brain freezes.

Let anxiety be and it will be less so. And quite possibly beautiful and exciting, too.

71. Of course, when we're anxious we're mostly told to calm down, to turn the dial down, to . . . *just relax*!!!! Are you like me and find this possibly the least helpful advice ever? (As I Instagram-memed once, "Never in the history of calming down has anyone calmed down from someone telling them to calm down.")

Turns out, in the 2013 Harvard University study, researchers concluded that for most people it takes less effort for the brain to jump from anxious feelings to excited ones, than it does to get from anxious to calm. In other words, it's easier to convince yourself to be excited than to bloody well *just relax* when you're anxious. Surprisingly, though, the excitement reappraisal didn't actually make the subjects less anxious, nor did it lower their heart rate. That's because the underlying anxiety was the same—it was just reframed as excitement.

72. To round things off, let's indulge in a list of reasons why anxious people are not that bad to have around. Our anxiety does have some beautiful kickbacks for those in our orbit. Again, you might like to leave this lying around open to this page. Or Instagram it. With a #justsayin tag. Subtly and humbly, of course.

Planning a picnic? Get an anxious mate on board—they'll be able to provide you with a full itinerary of weather contingency plans. And better salad delegation techniques. Jerome Kagan, who spent sixty years studying anxiety, says fretters are "likely to be the most thorough workers and the most attentive friends." This is nice to know, no?

Planning a dinner party/holiday/walk in the park/any kind of event in the next 365 days? Their phone will be charged, they'll have remembered Oliver is gluten-free, they'll have factored in dinner with your mom next month and your couples counseling appointment at 5pm.

They can spot a dickhead. Their heightened threat radar means they're selective about who they befriend. If you're one of their mates, you can rest easy knowing you're not a dickhead.

If you're about to be mugged in an alleyway, stick with the anxious person; they'll have all possible escape routes mapped out upon entry.

They'll hear, and attend to, the dripping tap well before you finally hear it and have to get up out of bed to turn it off.

They will book you both the quietest room in the hotel. They just will.

They're tough. As a functioning anxious person, they work hard as a matter of course. When the going gets tough, they naturally rise to the occasion. As I've said before, real-life threats are a cinch. Equally, they have tomes of great life advice when you're going through a rough patch. Veritable oracles.

They give a shit. About everything.

pain is important

73. Anxiety is painful. There's nothing quite like it. It's extremely private and lonely and it comes with the overwhelming sense that no one on the planet could possibly relate to the intensity and the sharpness. The whirling thoughts are so uniquely *you*, in a "stale bedsheets smell after a bout of the flu" kind of way. It's every thought you ever had, all at once. No one could ever understand so many thoughts. Which is why when someone asks me, "What's going on? What are you anxious about?" there is no way to explain.

And yet anxiety also feels like an original, human pain that we *all* just *know*. You sense this at its desperate depths. It can feel almost primordial. Edvard Munch's *The Scream* is anxious pain in oils. He painted it during a panic attack, it turns out. "I stood there trembling with anxiety—and I sensed an endless scream passing through nature." Yep, anxiety feels just like this for me, too. It emerges from a primal core and surges forth violently to outward expression.

The pain of anxiety is also unique in that there seems to

be no mechanism for its satiation built into our collective experience.

The pain of hunger makes you eat.

Thirst makes you drink.

Garden-variety fear will see you flee or fend or act or fix.

They all have a rather pleasant endpoint that makes the pain worthwhile. But anxiety doesn't direct us to its solution. Instead, as we've covered already, anxiety only seems to make us . . . more anxious.

So, yeah, anxiety is painful in lots of different ways. Can we move on now?

74. Just as I started writing this book I set off for a month to an Ayurvedic clinic in a muddy paddock in Coimbatore, India. (Ayurveda is an ancient Indian approach to wellness that makes profound, intuitive, sense. Yoga as we know it today, all meditation techniques and a lot of the dietary theory that I espouse come from this tradition, which is more than 5,000 years old [some say 10,000]. Buddhism stemmed from it 3,000 years ago.) My autoimmune disease kept flaring and I'd been told that the ancient Ayurvedic practice of panchakarma is a boon for Hashimoto's. So I booked myself in.

– If you want to learn more about Ayurveda, Deepak Chopra's book *Perfect Health* is a good start.

Entirely predictably, I overresearched "the most authentic Ayurvedic clinics in the world." I opted for the most austere route, rather than one of the more comfy "spa" set-ups. This route involves living at the clinic with mostly very sick locals where you're assigned a doctor and therapist who monitor you daily.

This kind of clinic doesn't pretend to be anything but grim. And, for me, the experience almost seemed tailored to press every one of my buttons. I wasn't allowed to bathe, move, internet toggle, or read. In the evening, a therapist bathed me in mung bean powder and tepid water as mosquitoes mauled me. There was no toilet paper or towels (we were given torn-up cloths to dab ourselves with). There's no TV, air-con, or daiquiris by the pool. We were also told to not think about sex. On this occasion, I found myself decidedly untempted.

I'd been at the clinic three weeks and I was in all kinds of physical and psychic pain. A monsoon had hit. Mosquitoes continued to devour me. Mold was growing up the walls. I lay every day on my wooden single bed, my unwashed hair wrapped in an old rag, staring at the ceiling fan. I wailed into the humidity.

One morning Dr. Ramadas, the specialist assigned to me, came and sat on my bed and we started discussing my tendency to flee pain, to ricochet into brash solutions and fixes when I get uncomfortable.

He smiled through his greasy glasses with his clear eyes. "Why do we all expect to be happy? We all came out of our mothers crying. Pain is what we do."

It reminded me of a tweet from Alain de Botton several years back that sparked a Twitter chat between the two of us: "Happiness is generally impossible for longer than fifteen minutes. We are the descendants of creatures who, above all else, worried."

Indeed. The great worriers of history were the ones who saw the charging rhinoceros first, had an action plan ready to go should a tiger invade camp, fretted that the basket of weeds collected that day may be poisonous. We carry this terror in our genes into our suburban lounge rooms, to our office water coolers, to our IKEA-issue bedrooms.

Worry is our default position.

I add to this: We humans are the only creatures on the planet who can't sleep even when we need or want to.

I also add this: We are the only creatures with the capacity—nay, propensity—to ponder our inevitable deaths.

True and painful stuff.

I read an Existentialism 101 critique of Jean Paul Sartre's book *Nausea* one night babysitting the neighbor's kids back in my late teens. It was some kind of high school reader I think. The ideas were something of a revelation. It had never occurred to me that others found the contemplation of one's existence painful. Years later I read the original book. I was further comforted (and went on to study French, then German, existentialism). The interesting thing is that back in the 1930s *Nausea* was celebrated as a wonderful expression of the essence of the human condition. Today the main character, Roquentin, a thirty-year-old loner who felt sickened by the realization that he lived in a world devoid of meaning, would be diagnosed with generalized anxiety disorder, and prescribed an antidepressant or invited to undergo a course of cognitive behavioral therapy.

75. Synchronicity happened and I met Brené Brown. There were three strikes, funnily enough. (I think I've already mentioned my three-strikes-and-I-act-rule.)

In one week, three people mentioned the University of Houston scientist who wrote *The Gift of Imperfection*. "You should read Brené Brown. And download her TED Talk 'The Power of Vulnerability.' More than 25 million others have. You'd love Brené Brown," they all said. Who was this damn Brené character? I Googled her and learned she's best known for intensively researching the nature of vulnerability to the point of breakdown, emerging with a dead-gutsy roadmap for living a "whole-hearted life." My kind of fretter!

So I contacted her to see if I could interview her for the better life–searching column I wrote. She replied within 2.378 minutes to say she was due in Australia. In three days. *Bam!*

When we met in Sydney, in a suburban town hall where she was doing a public talk later in the day, I told her about my three-strikes introduction to her work. "I do the same thing!" she cried. We found this very funny and then compared all the other weird stuff we do to comprehend life and anxiety. She says the Serenity Prayer a lot.

"And I do this thing where I twist a special spinner ring when I'm uncomfortable and repeat a mantra: 'Choose discomfort over resentment.' "

Freud believed anxiety attunes us not just to external threats (charging rhinos, dodgy people in alleyways, off milk) but to internal threats and the need for growth. The discomfort Brown mentions brings this growth perhaps. Anxiety is a sign we need to move and change our lives.

"You've got to just sit in it, sit in it, sit in it," Brown told me. (She has also stopped eating sugar and given up caffeine to help her deal with her anxiety.)

We stayed out the back of the hall, talking fast and holding each other's hands when we got to intimate subjects. We both cried. You know those tears of recognition? It was a Saturday afternoon and everything felt out of time, like when you walk out of a cinema in daylight.

76. How does one sit in anxious pain as a matter of course? I mean, we often hear this kind of thing said, but what does it actually look like on an average Tuesday when you have the life blahs and you really don't have any tolerance for anything that sounds like it comes with a pack of angel cards.

We can sit with it by talking to it. Hello there, old friend, you're a bit needy today. Tell me about it. Yep, you're rage-y today. You're lodged just under my solar plexus.

We can feel into the physical discomfort and find it interesting to observe. I watch the tension build in my jaw, in my neck (as though it wants to extend forward and bite at something) and in my right hip (my action leg that wants to jerk me away).

I let myself cry from the loneliness of it all.

We can watch ourselves as we try to drown out the discomfort with a handful of corn chips or chocolate or raw oats. We can acknowledge what we're doing. *I'm not a slovenly food addict. I'm just shoving food on top of my anxiety.*

We can let ourselves be wrong. My weakness is an inability to be wrong. I mean, don't get me wrong, I'm wrong often. But I've become really seductive at explaining it away before I can be seen in my wrongness.

I got called out on something (too personal to mention here, sorry) recently. It made me squirm a thousand squirms. Admitting guilt also meant letting a bunch of people down and looking a fool for doing so. A double shame-whack. I wanted desperately to bombard the confrontation with bombastic and diversionary reasoning. But I didn't. I sat in the awkwardness and eventually said, ""I got that so wrong. I'm very sorry." Then a lovely thing happened. The other person softened and simply said, "That's okay. I can see you're sorry."

This from my Australian editor Miriam: "Oh darlin', we all do this! We'd always rather be right than happy (except maybe Jesus)."

We can waste a bit of time. Oh goodness, this is just the *worst* for us anxious folk—the feeling that anxiety is wasting our lives away. *I should be doing productive things! I should be efficient! I shouldn't be lolling about all numb and stunned on the bedroom floor in a fetal position! Life is slipping by! Friends' kids are turning into full-grown humans and overtaking me!* No. Stop. Let the time pass with seemingly nothing productive happening. The anxiety is important. It means something is in fact happening.

And it *might* mean coming off medication. To see what happens. Peter Kramer contends in *Listening to Prozac* that when we take drugs we don't just medicate away anxiety, we medicate away our souls. It's a controversial call. But, again, I know anyone who's been on anxiety medication has had that cringey feeling at some point (every day?) that their drugs might be masking something important that really wants to express itself. My friend Joseph told me he came off Lexapro when he realized this. "The stress and the anxiety were warning signals from my brain and body, like the pain that makes you move your hand back from a fire. They were doing just what they were supposed to do: telling me that something was wrong [in Joseph's case, that he was pushing his work life too far, to the detriment of his wife and two sons], that I needed to change something. The meds were switching off the alarm." He came off medication, reduced his work hours and moved cities to have a better quality of life.

I checked at Christmas—he's still off them.

77. To sit in anxiety is to stay a little longer. A little longer. A little longer. And to see what happens. We experiment with it, curiously.

"Let's see what happens." My meditation teacher, Tim, says this. "Let's, as in let us, as in you, me and the workings of the universe, simply observe what happens if you don't fight it, if you just stay." I do this. I stay in the muck and the mire. I like the idea that it's not just me on my own doing this. It's all of "us."

It's not easy. I don't think anyone ever said staying in your anxiety would be.

78. I once found a torn, heavily underlined copy of Austrian psychiatrist Viktor Frankl's *Man's Search for Meaning* at a bus stop in Andalusia. It had been stolen from a library in 1976 going by the faded purple stamps on the inside cover. It's a glorious book. Frankl was a Holocaust survivor who spent three years in German concentration camps, some of it in hard labor. He wrote the small book in just nine days straight after being released from prison.

The book shares what Frankl observed to be the unique human quality that distinguished those who survived the camps from those who didn't. He concluded that the prisoners who emerged (barely) alive were those who nourished their inner lives. In fact, the more sensitive men in the camps—those you'd expect to crumble in such extreme adversity—were the better survivors. Why? By nature they tended to not resist the pain and instead went inward to draw on this "inner life" when things got really bad. And this is precisely what saved them.

In contrast, those who perished tended to rail against the circumstances, clutching outwardly, expecting external factors to shift. They got sucked in by the suckyness of the circumstances. This clutching at the external exhausted their reserves and provided no salve, no nourishment.

Frankl also concluded that the purpose of life is to suffer. Actually, he went further. The purpose of life is to suffer *well.* By which he meant to go down into pain, own it, and not run from it. To sit in it. And in the process find meaning.

To be specific, Frankl maintained that finding the meaning of life is our ultimate purpose and suffering brings us to this purpose.

This is the most important thing a human can do, he said, regardless of whether you're a prisoner in a concentration camp, a starving kid in Africa or a WASP despairing in banked-up traffic.

I read *Man's Search* on a hike through the Sierra Nevada. I'd just handed the manuscript for my first *I Quit Sugar* book to my publisher. I was still living in my army shed in the forest but had increasingly been spending time in Sydney, public speaking and setting up my online business. An invite had come through to join *National Geographic*'s Blue Zone team in Ikaria, Greece, to take part in their study of the lifestyle habits of the world's longest lived people. I'd e-met the chief explorer via the comments section on my blog. We became friends, debating life principles and sharing mountain biking stories. The invite was intriguing enough and I'd been itching for my next adventure. I stopped off in Spain—a last minute decision—en route.

It was a five-day hike through a rocky, barren landscape. I'd designed the trip such that each town was between six and nine hours apart on foot, with nothing in between. I set off on my own with a very worn Byron Bay Markets shopping satchel slung over my shoulder. I would eat two omelets and coffee for breakfast in what was often the only taverna in town and carried in my little market bag a toothbrush and toothpaste, earplugs, a bottle of water, a cucumber and an orange or a few tomatoes, a topographical map and my phone (used as a compass). And Frankl.

I saw no one. Some days I'd encounter a donkey. I talked to myself the whole way, thoughts unfurling, ideas drifting across my path, not needing to be grasped and recorded. Or shared across social media. It was hot, 104-degrees-Fahrenheit hot. And I was wearing unsupported running shoes. I got hungry and thirsty and lost and scared.

It was a good backdrop for reading Frankl and every afternoon I'd find a shady spot and read a chapter or two. Clearly my hike shared little with Frankl's experience in concentration camps. But the underlying message of his book, which he wanted to share beyond those who were imprisoned as barbarically as he was, is that suffering, no matter the degree, was something to sit in, not flee.

The grittiness of planting one foot in front of the other sank me down into a rarely experienced (in contemporary life) rawness. It was painful, and after about five hours, every bit of me wanted it to be over. You can resist this discomfort, find the flies unbearable, give in to the resentment, torture yourself with the impatient urge to get to water and the meal you've planned in your head. That bloody meal. You've repeated the components over and over and over.

Or you can bunker down and sit in it. And when you do, something happens. You enter a slipstream of movement and calm nonthinking. You have to. Or you'll throw up. Or you'll get a case of the *ugghhhs*. Seriously, have you been to this place, where any kind of conscious, unmindful, resistant thinking seriously makes you too nauseous to continue? Down into the is-ness of it all you must go if you want to make it to water and a meal.

When this happens on these long hikes that I do every six to twelve months I become part of the landscape. The thinking stops. Give it another hour or two and I become part of everything. A deep feeling of knowing, of fitting in to a flow, descends.

Hiking sees me realize the value of sitting in rather than resisting pain. So does anxiety. Both see me draw on my inner life and bring me in closer.

79. "Most people shoot for happiness but feel formed through suffering," wrote David Brooks in his recent book *The Road to Character*, which I happened to read (clandestinely) in my grim room in the clinic in Coimbatore.

Happy is fun, sure. But "rich" and "deep" light my fire so much more. I've never been a happy type. Personal pleasure as a primary pursuit has always seemed empty.

I feel awkward around happy things, like drumming circles and parties. Often it feels like something we seek in order to avoid suffering. If it's a by-product of an experience that brings me closer to meaning, then that's different.

It might seem ironic that all the studies on the matter wind up showing that those of us who prefer to delve into the meaning of life tend to be "happier" than those who don't. Seriously, there are countless studies to this effect. One University of Arizona psychologist who published one of the larger studies on the matter said: "We found this so interesting, because it could have as well gone the other way—in the sense of 'Don't worry, be happy,' we could have found that, as long as you surf on the shallow waters of life you'll be happy, but if you dive into its existential depths, you'll end up unhappy."

He proposed that delving produced the deeper happiness because human beings are driven to find and create meaning in their lives, and because we are social animals who want and need to connect with other people.

Australian social researcher and author of *The Good Life* Hugh Mackay is a vocal opponent of the *pursuit* of happiness as a life strategy.

> *The pursuit of happiness seems to me a really dangerous idea and has led to a contemporary disease in Western society, which is fear of sadness . . . I'd like just for a year to have a*

moratorium on the word "happiness" and to replace it with the word "wholeness." Ask yourself "is this contributing to my wholeness?" and if you're having a bad day, it is.

We need to remember this, we anxious types who find happiness a slippery sucker at times and often get accused of taking life too seriously. Delving and seeking our purpose—yes, that Something Else—is what cuts it for most of us.

do

the work

80. Louise Hay has in front of her three sausages that she's placed alongside a generous mound of scrambled eggs from the hotel buffet. Plus precisely three prunes. She eats this ensemble bite for bite. Sausage, egg, prune. Sausage, egg, prune. She holds my hand between mouthfuls as we chat. I like all of this.

I'd emailed Louise directly and asked if I could meet with her during her Australian speaking tour. I was living up north in the forest, scraping by doing these interviews with Big Name thinkers about how to make life better. It's funny. I spoke to dozens of these Big Names over the years. I'd email cold. Not one ever said no.

Hay is almost eighty-five when we meet. She had healed herself of cancer more than half her life ago, going on to live an extraordinary life working with AIDS sufferers in the '80s then writing *You Can Heal Your Life* in 1984. The defining self-help book went on to sell more than 50 million copies. She also owns Hay House Publishing, which represents most of the gurus in this realm: Deepak Chopra, Wayne Dyer, Doreen Virtue, Gabby Bernstein et al.

She's had her fair share of tough times. But her success and recovery from cancer didn't entail anything particularly out of the ordinary, she tells me. She just made a lot of little "right moves." I tell her about my build-the-muscle mantra and how I was healing myself by building up healthier and more satisfying life habits one small trick, or change, at a time. She pats my hand, all grandmotherly. "Yes, I just answer the phone and open the mail." Which is to say, she gets up each day and does the work. She chooses. Then she puts her head down and plugs away. With time, life takes her to where she needs to be.

"You're a forty-year overnight success story," I say.

Hay laughs. "Anyone who thinks they can heal without doing the work is missing the point," she tells me.

81. Now, we can all sit around and *talk* about this crazy journey. Yep, we can even read this book and get fired up by the *idea* of making our anxious beasts beautiful. Or we can roll up our sleeves and *do* the work.

Frankly, if you've got this far into the book, you have little choice. You've sold the farm and you're on your way. You *have* to do the work. And, this far in, you know there are no shortcuts. No one else can do it for you. To heal, we have to weave our own unique path through it all. Because it's a responsibility.

I explored this as far back as 2013 in a blog post:

> *In my experience, living with a wobbly mind is akin to being charged with carrying around a large, shallow bowl filled to the brim with water for the rest of your life. You have to tread super carefully so as not to slosh it all out.*

So you must learn to walk steadily and gently. And be super aware of every movement around you, ready to correct a little bit of off-balance-ness here, a tilt to the left there. This is just the way it is. Living this way requires vigilance and is about constant refinement. If you waiver and get unsteady, the water starts to slosh. And if you don't bring yourself back quick enough, the sloshing gathers momentum and, well, you lose it. Right? And, just to really drag out this metaphor, this means you then have to return to the source and fill it back up again. Which is tiring. So tiring.

And just to push it a touch further: to carry the bowl steadily means walking in a pretty straight line. Which means there will be scenarios and environments and people that simply are not conducive to your journey. They're too bumpy or jarring or wobbly. Or crooked. Me, I can't do late nights at bars and I struggle around people who live loud and fast. Don't get me near people on cocaine—their frenetic energy drags me way out to the bumpiest of tracks. I say this with all kinds of clean-living piety—cocaine is antithetical to anyone on the path to truth. I don't say the same about other drugs, apart from sugar.

I point out that keeping our bowl steady is a responsibility. We *must* work at it. We must vigilantly build our stability so that we can carry our full bowl without sloshing all over ourselves and, perhaps more importantly, our loved ones.

Clearly reflecting where I was at the time, I finished the post with:

Yes, I'm tired of spilling on my loved ones.

If you're like me (hypervigilant, overly responsible, responsive to a cause), you'll no doubt find this a purposeful call to arms. And, if you're like me, anything geared at helping others rather than ourselves is generally very energizing and clarifying. Yes? For you, too? Good. Now go do the work.

82. When there's a fire, we don't decide to dig around, ask a few questions of onlookers, to understand how it started. We call in the fire brigade. Quick sticks. We can apply this to anxiety. When we're in anxiety, particularly an anxiety spiral or panic attack, we must focus on coping. Once it's abated, though, that's when we have to do the work. We have to ask the questions. Plus, we have to build the resilience and courage and muscle with a whole lot of little right moves to ward off further fires. As Canadian spiritual author Danielle LaPorte shared with me over the phone just now in a chat we were having about this very notion, "You've got to get in front of it, be prepared." Of course, it's hard to do this when the fire's under control and the emergency has abated. We'd rather just get on with life. But as I often say in such instances, it's only hard; not impossible. Roll up your sleeves, focus, and "hard" is entirely doable. Besides, it's a responsibility to do the hard work if you're someone whose anxiety ignites regularly.

83. Gabby Bernstein is one of Louise Hay's authors. We met through a London friend, Louise, who'd done my 8-Week Program. Gabby quit sugar on her own three years ago with the goal of healing candida. Her mind got incredibly clear. This helped with her own healing work—both professional and personal—and she shared her story online. We became e-mates. Lou and I did too, meeting up in Primrose Hill a year after the multi-e-intros.

When I went to the U.S. to launch my first book, *I Quit Sugar*, Gabby and I arranged (via a series of rapid-fire emails) to meet for dinner. On arrival in New York I went to do the first of a dozen or so TV interviews. Suddenly I hear a high-pitched "Oh my God, Sarah Wilson is in the building!" In blew Gabby, all bangles and bejeweled silk. She was launching her book on the same day, on the same show, in the same timeslot, and on the same day we were meeting for dinner. She'd just seen the show's call sheet with my name on it. Gab just *luuuuurved* the synchronicity. I did, too.

We get to Angelica Kitchen, the vegan joint in the East Village she'd booked, our TV hair and makeup very out of place. She tells me that she's a control freak who can burn herself out. She's not outwardly Zen, but she keeps her keel aligned in her own way . . . with little right moves it would seem. She shares a bunch of her lifestyle habits, including praying and doing yoga. Foremost for her is "just showing up."

Ninety percent of success is about just turning up, she reckons. She feels crap some days, but she'll commit to showing up at her yoga class, turning on her computer in the morning, saying "yes" to a request. When you get that far (to the yoga mat, to the desk), you're most of the way there.

I totally agreed. I still do. Facing my anxiety in the way that I've chosen to has not been the easy path. Going my own way has been daunting at every step. I'm constantly tempted to reach outward again, to a new fix, a new guru. Gabby's New Yorker forthrightness lured me in, for instance. But I've found all I need to do is take the first step—commit, show up. And my path unfurls from there.

It's like when I have to motivate myself to do exercise when my inflammation, or my anxiety, flares and it hurts to even think about leaving the house. Gentle movement has been shown to help both. I know this. I have to do the work on this one, despite the pain entailed. I simply tie on my shoes and walk out the door and commit . . . albeit to a mere fifteen minutes around the block. But once I start walking, you see, I'll mostly—actually, always—find the pain and fogginess backs off and I want to go further. And with a bit more spring in my step. Showing up provides me with enough forward flow to keep things moving. You know, that's how it goes with most things.

I don't really get stressed about how I'll fare after a sleepless night, even when I have to do morning TV the next (or the same) day. I now know that so long as I turn up, it will work out from there. A twenty-minute meditation in the park on the way to the studio will get me into roughly the right mindset.

It was the same with quitting sugar for me. Back in January 2011, I merely committed to "giving it a go" for two weeks, rather than "forever." The low aim helped me to just show up. Once I got into it, though, I found it felt good and my skin changed—both my pimples and wrinkles faded (this happens in 10–14 days for most people). Perfectly and flowingly, this vain feedback loop compelled me to keep going and going. I would have balked, as would the millions who've since followed the same program, if I'd set out to just-goddamn-stop-eating-the-stuff forever.

Simply show up. Start. Things will flow.

84. And yes, it's hard. But as F. Scott Fitzgerald—a bloke who loved to self-torture—put it, "Nothing any good isn't hard."

And yes, going out on your own and doing this kind of work takes time. But nothing any good happens overnight, either. I emphasize this because I know that the time required to establish your own unique brand of healthy habits that allow you to live beautifully with your anxiety may put you off. A pill, or a new self-help book, promise to be so much quicker. And I get it. I'm incredibly impatient. I vibrate with impatience and so my own journey has entailed looking into the worth of taking a *loooong* time to do things. Must it? Are there shortcuts?

My research uncovered that Bruce Springsteen spent six months recording "Born to Run." Leonard Cohen took more than five years to write "Hallelujah," possibly one of the most perfect lyrical creations ever. Australian musician Paul Kelly spent seven years fiddling with "To Her Door." I spoke to multi-award-winning Australian author Kate Grenville recently. Her beautiful work *The Secret River* took eighteen drafts to complete.

As I learned about these creative tortoises, I realized that I, too, take way longer than most people to do pretty much anything that matters to me. It took me seven-and-a-half years to get an arts degree, two attempts to write this book (I got 60,000 words into the first one, five years ago, and tossed the lot), two years in the career wilderness after the magazine and TV experiences in my mid-thirties to feel right about what I was going to do next, and about eighteen months of running my sugar-quitting program for free before I felt comfortable about charging money for it. And this anxious journey? It's taken somewhere between fifteen and thirty years (depending on what starting point I take) of showing up, day after day, for me to arrive at a point where I can say I find it a beautiful thing.

85. When I lived up north in the forest, I used to drive to my friend Annie's house in the hills for dinner on Sundays.

I timed it to be able to listen to Ira Glass on *This American Life* on the radio. If you haven't listened to one of Ira's meandering, whimsical interviews about life, you truly should. He's a cerebral and emotional creative who takes the collective to places in our shared experience that we tend to avoid, or struggle to comprehend and articulate. Of course, I dug deeper into the guy's story and came upon this quote of his, shared in an interview. It's about creative work. I do feel, however, that the struggles that creativity presents are not dissimilar to those we, the anxious, face:

> *Nobody tells this to people who are beginners, I wish someone told me. All of us who do creative work, we get into it because we have good taste. But there is this gap. For the first couple years you make stuff, it's just not that good. It's trying to be good, it has potential, but it's not. But your taste, the thing that got you into the game, is still killer. And your taste is why your work disappoints you. A lot of people never get past this phase, they quit. Most people I know who do interesting, creative work went through years of this. We know our work doesn't have this special thing that we want it to have. We all go through this. And if you are just starting out or you are still in this phase, you gotta know it's normal and the most important thing you can do is do a lot of work . . . It is only by going through a volume of work that you will close that gap, and your work will be as good as your ambitions. And I took longer to figure out how to do this than anyone I've ever met. It's gonna take a while. It's normal to take a while. You've just gotta fight your way through.*

86. Anaïs Nin writes that anxiety can kill love. "It makes others feel as you might when a drowning man holds on to you.

You want to save him, but you know he will strangle you with his panic."

Ain't that the truth. I see that look on others' faces when I'm drowning in one of my spirals. I know that many of the loved ones I've turned to, or allowed in to witness me in this state, have had to swim away from me and look after themselves, leaving me to drown. I've always feared that they think I'm going to strangle them emotionally with my complexity. So I usually send them on myself.

Sometimes, though, when I put in the work, my anxiety has seen love grow, not die. And so, anxiety can be the very thing that pushes us to become our best person. When worked through, dug through, sat through, anxiety can get us vulnerable and raw and open. And oh so real.

My self-destructive inclination when anxious has been to distance myself from the world. I'm really very good at extracting myself from those around me and hiding out until I think I'm a more bearable person to be around. After my first panic attack in my late teens, and in response to my bulimia that had blown out of control, I took off to Europe and didn't surface again for a year. Living in that shed in the forest for eighteen months was a lot about removing myself from everyone. I've left new relationships, avoided relationships for years at a time and regularly announce I'll be working from home when I feel I'm too much for my staff. I convince myself I'm doing it for them.

But part of my journey has been to resist this fleeing. I've trained myself to tell partners and close friends and even relative strangers about my "stuff." At the time it's like standing naked in gale-force sleet. I tell new partners my ridiculous sleep routine, upfront. I feel so exposed and vulnerable—I mean, who tapes their lips together every night? But this is the point. Being vulnerable is the greatest gift you can give a loved one. Brené Brown tells us this. Being vulnerable is saying "I love you" first, it's doing something where there are no guarantees. It's being willing to invest in a relationship that may or may not work out. And it's staying to tell your truth. When you do, it provides a glorious space for a loved one—or a potential loved one—to step in and be *their* best person.

I tell my younger brothers and sister now about my various neuroses. I'm the capable big sister who always sorted out others' issues. But in recent years this dynamic has shifted. At Christmas now, when we pitch in to rent a run-down pine veneer shack down the coast instead of buying presents for each other, they wait until I arrive so I can choose the bedroom that will suit me best, the one with the least noise and with the least vibrations from the kitchen and the hot water tank. They don't make a fuss of it. And I adore them for it.

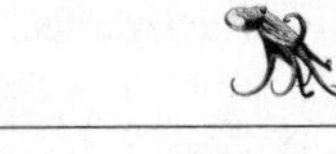

GET YOUR GUTS GOOD

Anxiety renders us delicate. Having your fight-or-flight response permanently switched to "on" triggers a whole stack of cortisol to circulate in your body which, among other things, down-regulates your digestive and reproductive systems. Hence, the high number of anxious women with PCOS and fertility issues, as well as gut-based complaints such as bloating, indigestion, heartburn, diarrhea and constipation . . . sometimes all at once. Elevated cortisol also causes poor absorption of key nutrients, particularly brain-essential ones such as the B-group vitamins, omega-3 fats, zinc, iron and magnesium.

Plus, anxiety makes us fat. And reduces bone density, leading to osteoporosis. Oh, and the inflammation associated with anxiety has a dastardly flow-on effect. The so-called "modern" metabolic diseases such as obesity, type 2 diabetes and heart disease are all about inflammation, which is why there's a higher incidence of these kind of diseases among the anxious.

Anxiety can cause all of these things. But it's a two-way street—bad health habits also cause anxiety. How do you modulate all this? Meditation works. So does walking. So does getting your gut in order. Specifically:

Quit sugar. When we're stressed we crave sugar (glucose is our body's preferred energy source), but consuming too much of it triggers a cascade of chemical reactions that promote chronic inflammation. Sugar also mucks terribly with gut bacteria balance which, as we have seen, plays a huge role in the health of our immune system and can directly affect mental health. So.

You really must reduce sugar in your diet—preferably down to the World Health Organization's recommended 6–9 teaspoons a day. I did, to combat my Hashimoto's disease.

Very recently, bipolar disorder has been linked to elevated uric acid levels. Sugar—specifically fructose—inhibits uric acid excretion and research by the University of Basel in Switzerland has found that a low-sugar diet improves symptoms in sufferers.

Just Eat Real Food (#JERF). This means pretty much anything that doesn't come in a packet. Which is much the same as saying "quit sugar" (when you quit sugar, you essentially quit processed food, since more than 80 percent of the processed stuff contains added sugar). A study published in the American

Journal of Psychiatry found that a diet consisting of vegetables, fruit, meat, fish and whole grains correlated with less anxiety. Another study found that combining junk food *and* stress (a pretty popular combo) was particularly explosive, particularly among women. When low-stress women ate junk food the health impact—across all markers—was a fraction of that for anxious women.

Eat 5–9 servings of vegetables and fruit a day. Some scientist some time back did an experiment that found anxious teens who ate extra servings of green vegetables saw a reduction in their anxiety levels. It's entirely possible to get to nine servings by starting every meal with a good three servings of veggies and fruit and adding protein and fat and carbs from there. I eat veggies for breakfast to hit the nine-servings goal.

For more of this kind of nutritional guff, head to — iquitsugar.com.

Eat yogurt and fermented stuff. I don't make and consume sauerkraut because it looks all quaint on social media. I do so because pretty conclusive evidence suggests that probiotic foods improve gut health.

Take some supplements. I'm not going to wade too heavily into this realm, except to say that you might want to see a good integrative doctor, nutritionist or naturopath with a sound understanding of endocrinology to get some tests done and advice on where you might need support. Many of us are deficient in magnesium, since our food is grown in magnesium-depleted soil. But evidence also shows it's a boon for calming us. Take an Epsom salt bath or try magnesium citrate or a topical magnesium gel. Vitamin D is worth looking into, ditto vitamin B6 and vitamin C. And zinc. And having your thyroid levels tested. I could go on. But it's best you go on this journey with someone qualified.

indecision

87. After I left *Cosmo*, and just as I launched headlong into my protracted Mid-thirties Meltdown, I suddenly had time to do what I'd seen other people doing in cafés at 11am when I was darting across town to client meetings, stressed and preoccupied. So in that suspended period where I was in denial about my disease and avoiding doctors, I set about spending each morning sitting in a café at inappropriate times doing the tabloid cryptic crossword and reading back issues of *The New Yorker*. This was the grand idea.

I'd set off at the languid hour of 10:30am from my flat in Bondi. But something would grip me as I approached the beachside café strip. Which café should I go to? I couldn't decide. There were too many choices. How should I decide? What was the best choice? The familiar momentum of thoughts would build. I'd do pros and cons lists for every option, then backtrack to try to work out what my gut felt. Then enter an adjacent dialogue as to the appropriate interplay between gut and head in making a decision. Then I'd become suddenly very anxious that I *should* be able to make such a simple decision with no fuss. Everyone else can. Then all of it would be overlaid with a hyperawareness of how First World-y all this was. Yeah, yeah, the starving kids in Africa again. This would all cluster and fester in my head and fast become a throbbing panic. I'd stop dead in my tracks, unable to move.

I tried to make it to a café on about a dozen occasions. But I'd wind up sinking to the gutter in a truly awful state of analysis paralysis.

On two or three occasions I managed to get myself seated at a café. It took playing fate games to achieve this. If it's an even-numbered time on the clock as I walk past the café on the corner, that's where I'll sit down. If not I move on down the strip. But I'd stall again. Long black or cappuccino? How do you decide such things? How do you have a preference? How do people know they like chocolate over vanilla? I just didn't know. I write this truly hoping you know what I mean and that you've had a similarly ludicrous decision-making experience.

My publisher contributes this in her comments in the first draft: "I remember a woman telling me about when she came back from being an aid worker in Africa. Emotionally, she'd held it together while away, but broke down in the undies section of David Jones deciding between boy cut and bikini."

I've done the same in toothpaste aisles in supermarkets. Spearmint or teeth whitening? I read that Jonah Lehrer, author of *How We Decide,* does exactly the same, despite being an expert on how to make decisions. So I phoned him immediately to find out why such banal decisions stall the anxious. He tells me it's because we allow ourselves to be fooled into thinking they're important decisions. "Call it the drugstore heuristic: We automatically think if there are lots of options presented that a choice must really matter, even if it doesn't." The guy landed in trouble for fabricating quotes from Bob Dylan (of all people!) and misquoting others in his books, but on this matter his thinking is backed by a number of reputable studies. The most famous was conducted by

Sheena Iyengar, a blind researcher whose *The Art of Choosing* left me obsessed with the topic. She conducted studies in supermarkets with jam sampling displays and found that consumers presented with too many options (twenty-four jam varieties) became paralyzed by the choice and failed to purchase; consumers faced with just six were ten times as likely to buy one.

Of course modern life is one big cluttered drugstore shelf. Choice is sold to us as providing freedom. It empowers us, says the consumerist model, to define who we are. Which we know is just the most absurd thing ever.

88. When you're anxious, decisions can be your undoing.

Anxious people are shocking decision makers.

Plus, the process of making decisions heightens anxiety.

Plus, any kind of indecision or faffing or vagueness around us tends to trigger anxiety. I've already touched on the difficulty of easygoing types. A flaky arrangement with someone not predisposed to everyday commitment ("Oh, I'll see what my Saturday looks like and get back to you") or making a team decision with someone who, all laid-back and cruisey, defers to me, proffering not a trace of input as to what movie we should see or which direction to head in upon arriving at the Piazza del Duomo, can send me over the edge. Yeah, we could put it down to my being a control freak and needing to have everything in place. But it's more that considering another's needs literally doubles the amount of swirling options.

To my mind, decision-making is a key trigger of anxiety. I posed this idea to the Black Dog and SANE Australia focus groups, and to my own network. The responses support my sweeping claim:

> *I get so flustered by being asked certain types of questions that my brain just seems to disappear and everything is just blank and frozen. Like whether to take time off of work because I'm really struggling at the moment just exhausts me to even think about and I end up so frozen that it's easier to keep going (even past breaking point) than it is to make a decision to change something.* – Lisa Jane

> *Not being able to make firm decisions makes you feel less manly and capable, which then makes you more anxious.* – Tim

> *Yes, totally for my husband. Anxiety grows with the amount of decisions he has to make, he can't plan anything. I try to get my hubby to make decisions, oh no, he just can't. He says it's because he hates to disappoint anyone. He's relieved when his wife (me) makes the decision for him.* – Shaz

Dearest Lisa Jane, Tim, Shaz and Shaz's husband, may I provide some comfort here? There's a reason decisions bring us undone. First: biology. When faced with options, our two decision-making-centers—the prehistoric limbic system (which makes impulsive choices) and the neocortex (which can look ahead to the future consequences of such choices)—are having a go-nowhere tug-of-war. If you're anxious, your neocortex tends to be particularly fired up, so the tug-of-war is much more aggressive.

Further, a relatively recent study reported in the *Proceedings of the National Academy of Sciences* found that the anxious tend to have decreased "neural inhibition," a process that sees one nerve cell suppress activity in another, which is critical in our ability to sift through choices and make decisions. The worse the anxiety, the less neural inhibition we have. Indeed, drugs that increase this cell suppression (thus helping with decision-making) are being used to treat some cases of anxiety. See, there's a scientific connect that can explain our challenge! What sweet relief!

It's also worth knowing that choice has historically been shown to present many an anxious soul with some of the gnarliest existential angst going around. Let's turn to old misery guts Kierkegaard again, who famously wrote, "There would be no anxiety without possibility." By which he steadfastly meant that even the smallest decisions open us up to the realization that the possibilities are limitless. When we see this limitlessness, we must also face, well, that it all ends soon enough in death, and that everything is pointless. The very real risk and probability of abandonment, aloneness, and being rejected and unloved are all wrapped up in this awareness. The everythingness and nothingness all at once. It certainly hurts the head.

Anxiety is the dizzying effect of freedom, of boundless possibility, he reckons. As humans we want all options. But we have to choose one thing over another from the boundless, or unlimited, options to create our identity. This is angst-filled by virtue of the enormity of the task and the perceived risk of failure—what if we get it wrong?! I studied German philosopher Martin Heidegger for six months during my bipolar period in my twenties. He saw anxiety as the awareness of the "impossibility of our possibilities." I was living it out as I read his riddle-ish work. Of course the Germans have a word for this torture:

> **Zerrissenheit:** *(noun)* disunity, separateness, inner conflict; an internal fragmenting or "torn-to-pieces-hood" from toggling so many choices.

So you see, my friends, your pain is actually very understandable. We must now be emboldened by this knowledge and find entirely cheeky ways to sidestep the gnarliness of choice. There are ways and means . . .

89. Apparently, Victor Hugo wrote nude. Which is not a pretty visual. The writer also used to instruct his butler to hide his clothes so he couldn't head outside for a wander when he was meant to be writing. This limited his options so that he could focus on what mattered. Which makes me think, if only I had a butler.

– Actually, I've had my assistant Jo's support for six years. She works tirelessly to strip me of options, bossily telling me where to be, what to prioritize. She knows this calms the waters.

It's the paradox of choice, a phrase coined by American psychologist Barry Schwartz in 2004, drawing on Sheena Iyengar's work with jams and the like. More choice is meant to bring us more freedom (so says capitalism). And yet we're happier when we're bound. In fact, to be rendered choice*less* is the ultimate freedom. Iyengar's studies looked at marriage in her parents' Indian culture, comparing couples who married for love and those in arranged marriages. At the start of the study, the former are happier, but when the same couples are revisited ten and twenty years later the arranged marriage couples are close to twice as happy.

During that period when I finally gave up running off to New York to write about porn stars and to Peru to scale mountains and came home to my Hashimoto's diagnosis, I had my mobile phone, my old but much-loved Toyota Hilux Surf (along with my surfboard, signed by Layne Beachley, which I stored in the back) and my equally loved mountain bike stolen.

Then my computer died. Stopped. Dead. In the middle of the night. This all happened in the space of three days. Oh, and the phone service in my street—and *only* my street—went down. For two weeks. Which made it nigh impossible to sort out my stolen goods situation. I was also, if you recall, too weak to even walk. It took me three weeks to find out my car insurance had lapsed two weeks before the theft. And on and on went the ludicrous misfortunes.

I'd started seeing my meditation teacher, Tim, by this stage. He laughed when I rang him from a pay phone at the end of the street and told him my predicament. "You've been rendered choiceless. How good is that!"

"Not very," I said, somewhat indignantly.

It's just occurred to me that I don't think I've mentioned that during this same nebulous period in my life, during which I frantically tried to fight my exhaustion and illness and dwindling career relevance, while going steadily downhill (a period that lasted a good twelve months), I also took to adrenaline-based experimenting. I tried out sky-diving, Formula One racing and wake-boarding and took off to the Solomon Islands to write a feature about diving with sharks. All of which was clearly about grasping outward, away from myself and everything I was truly looking for. I mention such experimenting because it also concluded in disaster. While I was on the islands, all three planes—all three!—that service the archipelago broke down. I was stranded for five days on an island that was in the middle of tribal infighting and had no internet. I can see now why Tim found it all pretty funny.

Actually, I soon grasped what he meant by being rendered choiceless and why this is such a glorious thing when it happens. I only had one choice available. To stay put. To give up fixing and meddling and grasping outward. I'd missed the previous memos that tried to tell me this. They were subtle at first. But now the message was comically clear: *You cannot move. You cannot communicate. All that's left now is to stop, rest . . . and come in close. You have no choice.*

When I saw things in this very obvious light, well, I stopped. At least for a while. It was incredibly freeing at the time. When choicelessness strikes me now, which is to say, when it hits me over the head with its obviousness, I remind myself, "How good is that!" and give in with palpable relief to being told what to do by the funny circumstances.

You'll be glad to know, though, there are other, less dramatic, ways to render yourself choiceless.

90. When I was writing the Sunday newspaper magazine column where I spoke to famous and/or highly successful folk about how they make their lives better, I noticed that many of them ate boring breakfasts. Richard Branson eats fruit salad and muesli. Every. Day. Leo Babauta, the guy behind the very popular Zen Habits blog, eats Ezekiel flourless cereal with soy milk. Which just makes my eyes glaze over.

Many of them also wear the same clothes every day. President Obama wore the same style of suit every single day during his time in office. He once told *Vanity Fair*, "You'll see I wear only gray or blue suits. I'm trying to pare down decisions. I don't want to make decisions about what I'm eating or wearing. Because I have too many other decisions to make." Facebook's Mark Zuckerberg owns twenty versions of the same gray T-shirt and has said that what he wears each day, along with what he eats for breakfast, is a "silly" decision he doesn't want to spend too long making. Einstein reportedly bought several variations of the same gray suit for the same reason. Ditto Steve Jobs, who stuck to a black-turtleneck-with-jeans-and-sneakers get-up. As Henry David Thoreau once wrote: "Our life is frittered away by detail. . . . Simplify, simplify."

In his book *Uncertainty*, Jonathan Fields reports on studies he conducted with hundreds of successful creatives to discover what they did differently. He was looking for "the thing" that distinguished them from the rest of us. "Happy, successful entrepreneurs ritualize everything in their lives *but* their creative work," he wrote.

After reading this, I figured I should ring him to ask if he'd noticed the boring breakfast thing, too. He had. He explained that at every turn, successful creatives tend to streamline the minutiae of their lives so as to take out as much unpredictability as possible. "Breakfast choice and where you should have your coffee in the morning is minutiae, it gets in the way and saps decision-making energy," he tells me.

I'm someone who struggles with ritual and banality. I gravitate to novelty and can't bear to walk the same route to work twice in a week. But given my unimpressive history with choosing a café for my morning coffee, I started to settle into the idea of having a "regular haunt." At my regular haunts around the world (as I shift from city to city, unable to stay put), I noticed that they were filled with other nomadic regulars. The waitstaff all know our orders (mine: long black, in a glass, not a cup, with a side of hot water, for some "special" neurotic kick) and we are free to get on with our work. We look up from our laptops and notebooks and nod to each other every now and then. This is the second book I've written from my regular haunt in inner city Sydney. Other regular haunts included public libraries in New York, London, the Northern Beaches and Kings Cross, and bookshops around the world (the books provide a lovely, comforting "insulation"). I've become part of a worldwide community of minutiae-simplifying folk who are saving their amygdala muscle for better things.

Again, I don't think I'm making too drastic a leap when I say I think anxious folk are not unlike creatives (if they're not already both) in needing to reduce the number of choices they have to make so that they can fly free.

91. Behavioral psychologists refer to this technique as "dropping certainty anchors." Drop as many as you can to hold you firmly so that you can flap about as creatively—or anxiously—as required, like one of those inflatable, fan-operated men propped outside used car yards that jerk about moronically. The flapping about is manageable—and creatively productive—if we know we're not going to fly away.

We must drop certainty anchors. And I'm putting this in the imperative tense so that it's one less decision you have to make. You're doing it. No ifs or buts. And, apart from anything else, the world needs more certainty. Much of how we're living feels untethered and wobbly. What a wonderful thing to be on a bit of solid ground amid the flapping.

92. Behavioral psychologists also like to refer to decision fatigue. They liken our decision-making abilities to flexing a muscle. With each decision we make, regardless of whether it's big or small, we fatigue the muscle.

A study published by the *National Academy of Sciences* looked at the decisions of parole board judges and found that judges were significantly more likely to grant parole earlier when making a judgment in the morning. Cases that came before judges at the end of long sessions were much more likely to be denied. The rationale was that their muscle bottomed out from so much choice deciphering. Decisions become harder, more stressful and more confused the more decisions we make. The antidote is to automate as much as we can.

HAVE A MORNING ROUTINE

When I was interviewing all those successful life-bettering folk I would always ask, at the end of the interview, for their favorite hack, the one they personally live by. In all but a few cases they'd reply, "I have a morning routine."

Having a morning routine is a certainty anchor with really sturdy stakes.

Louise Hay told me when we met, "The first hour of your day is crucial." She starts by thanking her bed for the sleep (!), stretches, has tea, then goes back to bed to read. Because she likes it. She even made a great headboard so she can be at the best angle to read.

Leo Babauta's is to drink water, reflect on a cushion briefly, read something inspiring for half an hour, then write. "Before I check email or Twitter or read my feeds, I sit down and write. It doesn't matter what—a chapter for my new book, a blog post, answers to an interview someone emailed me, anything. I just write, without distractions."

Similarly, Timothy Ferriss (U.S. author of the mega-selling *The 4-Hour Workweek*) starts every day by journaling—he free-associates for a few pages in a notebook, "to trap my anxieties on paper so I can be more productive with less stress throughout the day."

Benjamin Franklin always woke at 5am to "rise, wash and address 'powerful goodness' contrive the day's business and take the resolution of the day; prosecute the present study; and breakfast." Each morning he'd ask himself: "What good shall I do this day?" Which is an adage I've adopted at the end of my meditation practice.

Stephen King keeps to a strict routine each day, starting the morning with a cup of tea or water and his vitamins. King sits down to work between 8 and 8:30 in the same seat with his papers arranged on his desk in the same way.

If you're anxious, you have to have a morning routine. Again, no ifs or buts.

Start off by letting go of the idea that you don't have time. Get up half an hour earlier and commit, in the first instance, to thirty minutes only. Show up! Do the work!

I rise at 6am. Non-negotiable. And drink hot water and lemon. I attend to ablutions. Then I slide straight into sneakers and move. I keep a bucket by my door with one pair of running shoes, one sports bra, one pair of green shorts. Every day (non-negotiable) I put on my one outfit (no room for faffing over what I'll wear) and get out the door. I'll decide what form of exercise I'll do the night before, so that I don't have to decide in the morning—a simple thirty-minute walk is perfect, a yoga class, an ocean swim, a surf—based on the weather forecast and class timetables.

I keep it lo-fi, doable. None of this driving to the gym or running park business. None of this multiple bits of equipment and drink bottles palaver. It only invites procrastination or piking. And more, not fewer, decisions.

Then I meditate. By exercising first, I find my mind is settled, ready for meditation. I meditate in the sun. A vitamin D hit to boost the package. If it's raining, I meditate inside. Decision made.

This, all up, takes about an hour, longer if I go to a yoga class. Then I shower and start my day, mostly with creative work in a regular haunt. As William Blake wrote, "Think in the morning, act in the noon."

If you eat breakfast, you might want to try eating a boring breakfast. Prepare it the night before so you know what you're eating. *Bam*, another decision eliminated.

I've been doing my routine for six years now. Every. Single. Day. When I travel, I adapt it. I meditate in the cab on the way to the airport. I run the fire escape stairs in the hotel. I establish a regular haunt for my coffee stop-off upon landing in a foreign country. I Google "best cafés near me" or I look out for fellow nomadic types looking over the top of their laptops for connection with like-minded strangers.

Once the certainty anchors are in place, the day starts and all kinds of chaotic decision-making can then ensue.

93. A few more certainty anchors that ritualize all that minutiae deciphering:

> *At the supermarket I buy the same brand of toilet paper, frozen peas and tinned tuna.* Ditto toothpaste. Heck, I buy in bulk.
>
> *I "shop like a man."* I find a style, a brand, a size that works and I buy the same thing over and over, only replacing things when they're threadbare. I've worn the same make of Target seamless undies for eight years now. I own seven pairs. I own minimal clothing—enough to fit in two suitcases. Narrowed choices means fewer "fash attacks" in the morning. Gray

drawstring pants, t-back bra and white singlet it is again, then!

I say yes. I play this game "I say yes" to anyone who comes to me with a solid, certain, already-decided preference. *Sarah, I'd really like you to join me for steak and a wine at the pub tonight at 7pm.* Yes, see you there. *I really want to watch a romantic comedy starring Jennifer Aniston playing a florist tonight.* Yes, suits me. This creates certainty. It means I gravitate toward certain, anchored people. My friend Ali is a great anchor. She extends deliberate invites on defined days. And emails a calendar request. I can't tell you how much I love a calendar request. I'm compelled to say yes.

I get recommendations. I learned this from someone at some point. I don't think it matters who. When making a decision about, say, the best TV to buy or the best hotel in Albury-Wodonga, ask someone who's already made the same decision—and has sifted through options—to tell you. Obviously someone whose research skills you trust is best. To this end, don't shy from online forums. The kind of people who go online to answer stranger's questions as to "the best cough mixture for insomniacs" are generally the kind of people who care a damn about most things. In a TED Talk titled "Why we make bad decisions," psychologist Daniel Gilbert explains that cognitive biases mean we are actually really bad at making decisions for ourselves and that even a complete stranger will increase the likelihood of making the "happiest" decision for you by a factor of two.

94. You can also work around the decision-making palaver by playing games to render yourself choiceless. If we're hell-bent on minimizing decisions, why not give in to the natural flow of things as indicated by the persistence of a particular "sign" or another's enthusiasm? As I've mentioned, I sometimes take to simply saying "yes" when someone issues a deliberate, decisive invite, and I will allow myself to be steered by "three strikes." In most circumstances it delivers me a great outcome and always saves me from decision anxiety.

Dad used to do this thing in the lead-up to the summer holidays that would decide our camping destination. He'd get The Youngest to close his/her eyes and stab at a map of the surrounding region.

"Right kids, it's decided, we're off to . . . Dubbo!" Oh, yes, the thrill! (No disrespect, mind, to Dubbo.)

Of course, the chances of stabbing directly at El Caballo Blanco fun park were slim. And so our annual camping trips were to places few families ventured for fun. Like Cowra. And Warrnambool. Although Miriam, my Australian editor, does inform me there is now an amazing adventure playground in the former port town. My memories from twenty-five years ago aren't so dazzling. One year we wound up at a goat farm outside Jindabyne during a crippling drought. The winds and dust forced us to dine each night with the recently divorced father who owned the farm and he served us microwaved meals. This was the only thing I remember from the trip. I'd never seen a microwave before.

"Really, what difference does it make?" Dad philosophized, folding up the map. He was right. Wherever my little brother stabbed, we always wound up in a campground with an above-ground kidney-shaped pool and ate goulash that Mom made in advance and carried in the stinking hot car in a big cast-iron pot and that we'd eat sitting on milk crates around the kero lamp.

FLIP A COIN. GO ON.

Like me, you might find the kind of caution-to-wind approach my dad used to take intimidating. You might also find the idea of "going with your gut" out of your league. An anxious person's gut is a fluttery mess; we don't know what we want. We are all head, little gut instinct. That said, I do have one trick that I play with a bit. It accesses a sneaky portal to the gut.

I flip a coin. But before I uncover it, I monitor my emotions to see what I'm *hoping* the result will be. There it is, my gut decision, peeking through my head clutter. This technique tricks you into thinking some divine intervention is going to make the decision and you switch to responding to the possible outcome. This switches off the decision-making muscle.

The challenge is to not then check to see if the coin flipped head or tails. Oh, and to use this moment to absorb the fact that it doesn't matter anyway.

95. Or you can . . . Just. Decide. Sounds frightening, but read on.

Back in my *Cosmopolitan* days I went to my publisher with two cover options for the following month's issue: Angelina in a green dress or Angelina in a black dress. Pat was a good decision maker. Or at least she made fast decisions that didn't get her worked up. On this particular visit I asked her how she did it. "If we're actually debating the two covers, going back and forth, then it means both are good options. Right? If one was really bad, you'd know about it." Yes, I can vouch for this.

"So I just choose one. It doesn't matter which."

Pat didn't break her rationale down, but I have many times in the retelling of the anecdote. On a two-party preferred basis, the black and green dress are likely to come in at roughly fifty-fifty. Sure, if we drilled down with a bunch of focus groups, the black dress option, for example, might come out marginally more popular—with, say, 54 percent preferring it.

But that's not the point.

What's important about making a decision is the "just deciding" bit. Because once you choose one—say, the black dress—you *make it* the best choice.

Pat did, in fact, choose the black dress. I returned to the office and told everyone the decision. "We love the black dress." The art team, mostly relieved to have the decision made, then created the best design and color format around this chosen image. The art director had seen a black dress done with an aqua background and a fluorescent orange masthead before. She was pumped. The subbing team worked cover lines enthusiastically. The whole office was loving the aqua treatment. Bit by bit the various departments massaged Angelina-in-a-black-dress to become a standout cover, the kind that gave me a satisfyingly *quenched* feeling when I walked past it on a newsstand.

There is never a perfect decision. They become perfect when we make them.

Louis C.K. also grapples with the descent into despair that decision-making can induce. I hate to give the disgraced comedian any further oxygen in this book, however, he has a 70 percent rule I rather like. If a decision—about a thing or a person—feels 70 percent right, he just goes with it; 70 percent is enough:

> *'Cause here's what happens. The fact that other options go away* immediately *brings your choice to 80. Because the pain of deciding is over.*
>
> And *when you get to 80 percent, you* work. *You apply your knowledge, and that gets you to 85 percent! And the thing itself,* especially *if it's a human being, will always reveal itself—100 percent of the time!—to be more than you thought. And* that *will get you to 90 percent. After that, you're stuck at 90, but who the fuck do you think you are, a* god? *You got to 90 percent? It's incredible!*

The funny thing is that behavioral studies show that we *think* making a decision is more anxiety-riddled than not making a decision. But, in fact, the opposite is true. The studies show that when we decide to do something and it turns out badly, it mostly doesn't haunt us down the track. We humans are master justifiers. Failing to act on a decision, however, will haunt us. The infinite possibilities of what might have been get us into all kinds of anxious messes.

If you think about this, it becomes apparent that we get anxious having to decide, but we also get anxious when we don't decide. If we know this, we might as well just decide. Right?

I share all of this, mostly, to lessen the potency of one choice over another. If we've investigated the options enough, it doesn't matter. Moving up, up and away from the chaos of indecision does.

back

the fuck

off

96. Can I just ask, would you describe yourself as a control freak? Would you describe your grip on life as white-knuckled? Do you tend to be even a little bit of a perfectionist?

I'll go first. It's a yes, yes and yes from me.

We, the anxious, fret. We meddle. We care (yes, we care!). We try to solve, to fix and to find the endpoint. This is what we do. Agree?

I said earlier that making decisions is a key anxiety trigger. If we drill down a bit we can see that this happens because we work to the belief there's a perfect decision out there to be made. But such a thing doesn't exist. And clutching at something that doesn't exist is enough to send anyone into a drowning panic. We can *never* find the best option. Anxiety is what occurs when we realize this, when we realize that we are not the captains of our own ships. What do we do next? We grip more, grasp outward more, get busier and more controlling.

I convince myself that controlling my life and aiming for perfection will cocoon me from anxiety. But it only causes more of the dreaded thing. — **cruel irony #15**

The Germans, once again, have a word for such heavenly clusterfuckiness (they really do understand the complex paradoxes of the anxious condition, don't they?).

Verschlimmbessern: *(verb)* To make something worse in the very act of trying to improve it.

97. We all just need to back the fuck off.

98. Back in Year 8 English Mrs. Cochrane set our class a metaphor assignment. We'd covered analogies the week before.

As ever, I took the task incredibly seriously. "There's a river that flows," I wrote somewhat loftily. This is life's inevitable logic, the logic that ensures eyebrow hairs sprout exactly where we need them to protect our eyes from dust, and ensures that springtime rolls around just as winter becomes unbearable, and that sees specks of moisture coalesce to form rainbows. I used these very examples in my assignment, but I'm saving you the tortured prose by paraphrasing things heavily.

Some of us try to dam the river with piles of logs and other obstacles because we think the river should flow differently, by micromanaging our partners or blocking our pain or by forcing a dinner that no one wants (they repeatedly cancel but we ignore the signs and keep rescheduling). When we do this, the pressure builds. And builds.

The water (the flow of life) banks up behind the obstruction, determined to continue its flow because, you know what? It kind of knows where it's going. It's ingrained in the groove of the valley, the gaps in the boulders, and it's bigger than us. Way bigger and way more knowing.

Eventually the flow wins out and *Boof!* our micromanaged pile of logs explodes from the force of the flow. Our stuff goes flying in all directions. It's devastating.

And, then . . . the river goes back to flowing as it was always going to. Before we came along and got in the way.

I round off my metaphor assignment (cringefully) by advising the reader (poor Mrs. Cochrane) to perhaps try using the logs to build a comfortable little raft instead and to sit atop it and let the river carry them languidly down the river.

I got a B for the project. Twenty-five years later Mrs. Cochrane connects with me over Facebook and we meet for meatballs in New York's Soho where she now works. I took my below-par grade up with her. I'd been really happy with my torturous metaphorical tale; I had wanted to take it up with her back then, but our school was not the kind that encouraged such proactive and empowered behavior. Mrs. Cochrane and I spent the rest of the night talking about the flow and the synchronicity of life, the very flow that brought us together that night. The very flow that my perfectionist thirteen-year-old self had wanted to interrupt by disputing her grade.

99. More flow. The next day I was reading Simone de Beauvoir's *The Blood of Others* at McNally Jackson, my favorite bookshop in New York. Flowingly. On page 108 (a mathematically perfect number) I find this line: "I wanted all human life to be pure, transparent freedom; and I found myself existing in other people's lives as a solid obstacle."

108 is an important number for me. I chat about why on page 297.

Yes! More than anything we want to exist in pure, flowy freedom. We don't actually want to build dams.

We need to get out of the way.

We need to let go.

We need to take our "filthy mitts" off life . . . for ourselves. And for others.

Besides, we don't want to feel we have to do it all on our own. I want to know that life "has this one." That I'm held and can chill out on top deck for a bit with a drink with an umbrella in it.

FOLD FORWARD AND SURRENDER

You might like to try this thing I was taught in a yoga class about ten years ago to access this flow business. At the end of the yoga class we would sit cross-legged in meditation for a few minutes. After five minutes the teacher invited us, gently, to fold forward from our hips over our legs and "surrender to the day," our arms outstretched in front. That's it.

Please do try it.

For there's something about collapsing into the earth like this that truly sees one "give in." It's not a novel idea. Supplication to the feet of the sublime has been doing the rounds for a long time.

You don't have to do the yoga and meditation bit first, although both open the body and mind perfectly. A walk to a park and a little sit on grass in the sun for a bit would no doubt do the trick. You might be holding a stack of tension and worry. The walk and quiet sit will soften this a little. But when you fold forward, notice if there's not a discernible release that kicks in.

Emotional tension is held in the hips and groin according to yogic tradition. This is stretched and massaged by the release. Feel into this. There's a "giving in" entailed in the gesture. Folded over, the blood running to my head and heart, I'm aware the day will run as it needs to and I just need to flow with it. I board the log raft.

The part I loved most was when the yoga teacher invited us all to rise to sitting and to "take it with you into your day." Try this bit, too. Try holding the feeling as you walk back home, as you make your breakfast, as you kiss your kids goodbye, as you sit down to write your first email of the day. See how long you can hold the languidness.

100. When I was nineteen I was mugged in Nice after hitchhiking to the French border from Florence. I was left with no backpack, no passport, no identification, no money, no credit card and with "only the clothes on my back"—a pair of jeans, black boots and a navy Sportscraft turtleneck (which I still wear).

You know what? It was one of the best experiences I've ever had.

After spending a day on the beach in my underwear with some U.S. soldiers who bought me lunch at McDonald's, I jumped the overnight train to Paris. I chose first class. (If I was going to get busted for jumping a train, it might as well be a decent ride.) Around midnight I was caught. I had no identification, of course, and insanely the guard believed my story and told me to get some sleep. He patted my shoulder.

In Paris the Australian Embassy would not issue me a new passport without $200. The bank wouldn't issue me $200 without a passport. You see my bind? This was before email and internet and mobile phones. It took two weeks for the issue to be sorted as I jumped the Metro back and forth between the bank and the Embassy.

For two weeks I was itinerant and renegade and it was one of the few times in my life I've not been anxious.

Which I know seems counterintuitive because I was neck-deep in everything that freaked me out—sleeping arrangements I couldn't micromanage, meals I couldn't preplan, and uncertainty in every direction.

I snuck into youth hostels each night and took whatever bunk bed I could find and I stole food from supermarkets. Old men at fruit stalls gave me peaches and baguettes when I walked past, perhaps picking up on my vulnerability. Peaches and baguette, then, it was for dinner. I would sit in the Jardin du Luxembourg and eat it, as happy as a girl can be.

I roamed and killed time and I would enter shops and feel the lightness of knowing I didn't have to make decisions as to whether to buy the cute T-shirt as a memento of my Parisian stay. And I couldn't take photos.

I had no control. No agency. I was a passenger for two weeks. My anxiety had nothing to grip on to.

Plane turbulence can produce the same sense of lightness in me. I'm not in charge, there's nothing I can do. So I sit back and enjoy witnessing the ride. Ditto my few stints in the hospital for various surgeries. I'm aware I can't do anything to help the situation and that my filthy mitts can't grasp at anything and so my anxiety goes on a little holiday.

In a similar vein, studies have shown that Brits who had neurotic disorders prior to World War Two experienced a decline in anxiety in the wake of bombing raids. A German study in the 1950s found the same among concentration camp prisoners with the same disorders—their symptoms went into remission during the war.

Like I've said before, emergencies put us into the present. And anxiety struggles to exist in the present. The hyper-attention required in a real-life upset flips us out of our funk. We can't be focused on our grandmother dying or a loved one's miscarriage *and* be anxious. They're two competing parts of the brain and the part that deals with real life wins out.

As I've flagged already, old anxiety happens to be a bit of a mono-tasker. You can't be anxious *and* be in the present. And you can't be anxious *and* attend to genuine fear or catastrophes. And you can't be anxious *and* walk mindfully or breathe deeply.

Let's pause on this for a sec. Let it sink in.

No need to spell it out further, right? Okay, then . . .

101. Even being forced—as in, absolutely beyond our control—head-on into our neurotic fears can abate anxiety. Matt Haig tells how his agoraphobia abated when he was thrust into a trip to Paris (coincidentally enough). His long-suffering girlfriend had bought him the trip for his birthday. Haig had previously been unable to go to the corner shop for milk without having a panic attack. But being forced beyond his fear threshold—and surviving—brought a certain relief. "The best way to beat a monster is to find a scarier one," he writes.

My friends have a young daughter riddled with anxiety and OCD. When she was eight or so, they told her she didn't have to go to her friend's party (which she was dreading) if she cleaned her teeth only once that night. The trade worked. The next day the poor little chicken excitedly told her mom, "Keep making me not do things, please." Keep rendering me choiceless, was her desperate plea! Bind me! This story still sends a chill through me. A child of eight had worked out the strange machinations of her anxious mind in a way few doctors probably could.

My own experience matches up. Unfamiliar surroundings can buck me out of my anxious rut, again because the usual stuff I grip on to isn't immediately apparent or available. I can often sleep on a train or a plane better than I can in a bed. You might think the vibrations and bright lights and snoring businessmen who've had too much complementary whiskey—some of my worst triggers—would send me over the edge. But it's almost like it's all *so* bad (a scarier monster), and *so* uncontrollable, I have to give in to the "is-ness" of it all. I'm rendered choiceless. I think, in some ways, this is why I live so nomadically. It keeps me in a perpetual state of being confronted with scarier, uncontrollable monsters. The adjusting and fending stops me from noticing the dripping tap three apartment blocks away. Knowing I'm leaving in two months' time means the waft of aftershave from the flat adjacent to mine at 7am every morning won't tip me over. I wonder if this is why so many anxious types in history lived itinerantly?

I read in a *New Yorker* article about some recent science that postulates that psychedelics may be good for sufferers of OCD. The drugs were shown to shut down the default mode of the brain and disrupt the repetitive and control-focused patterns of thought and behavior. "It may be that some brains could benefit from a little less order." Oh, my goodness, yes.

– My Australian editor, Miriam, had this to say about the theory: "This article just blew the top of my head off and I had to have a lie down . . ."

102. Let me touch on the myth of life balance for a moment. Some time in the last few decades we got it into our rigid heads that what we're really missing is a magical balance between time spent on obligations and downtime. Apparently, our lives are completely out of whack—we're pushed and pulled, and we just can't get the damn ratio right. If we could perfect this magic ratio *then* we could address our anxiety. Familiar?

I looked into life balance for quite some time. It became yet another obsession of mine. Was there a magic ratio? What did the research show? Oddly, not what I thought.

Basically, the science shows that unhappiness among women correlates with having more options. Which probably shouldn't come as a massive surprise to you this far in. "Having it all"—career, kids, access to the rowing machine at the gym—has come with the pressure of feeling that we have to "do it all." Women have got it into their heads that they *should be able to* do it all. And in perfect balance. And this has resulted in more stress and less happiness. (I speculate that men are feeling the same but it's just not reflected in the research yet.)

In response to these findings, U.K. pop-trend researcher Marcus Buckingham, who I did a bit of work with during my time at *Cosmo*, took a different tack and investigated, inversely, what the *happiest* women were doing differently. And his conclusion was this: they strove for imbalance. Messy, all-over-the-shop imbalance. This was new. And I liked the sound of it.

These happy women, he said, realized that balance was impossible to achieve and trying to do so caused unnecessary anxiety. I mean, how do you get the perfect ratio? For every new commitment you take on, do you allocate the same amount of time for sitting in a bath or doing a Meaningful Craft Project with your son? If a work project or sick partner suddenly requires more of your time, do you hold up the hand and say, "*Whoa, world! No can do! I'm behind on my yoga class quotient!*"

Instead, Buckingham found these chilled, happy women "tilted" toward activities and commitments that they liked and found meaningful. Amid the chaos. They didn't wait for the chaos and the commitments to get under control.

I love this idea. *Tilting*. It's when you have so much to do and you could list it all out and try to prioritize. Or you could just sit in the everythingness and lean toward stuff as it arises that feels right. Tilting doesn't involve holding up the hand and plonking a lump of logs in the flow. Nope. When you tilt you grab a log that looks about right and jump on.

103. Learning to back the fuck off is really hard if you've been anxious all your life. So you need to mess things up a bit.

In *The Happiness Project* Gretchen Rubin rattles off a number of rather rigid rules for having a better life, but places the qualifier: Every now and then do the opposite of everything she's just said. *Seinfeld*'s George Costanza inspiringly suggested the same with his opposite theory. He reasoned that since everything he'd done in life had been so wrong, if he did the exact opposite of what he normally did, he'd get it right. Right? It works; he picks up a hot woman instantly with the line, "I'm unemployed and I live with my parents."

Indian philosopher Guru Dev says the same: "Do the opposite of what you'd normally do." Why? It injects freshness. The jolt of going against the grain gets you to look at things differently.

Doing these things makes me edgy at the time. But I play with it, treating it as an experiment, because I know that doing so quashes Perfect Moment Syndrome, a term I think one of my staff made up when I worked in women's magazines, mostly for the resulting acronym. PMS afflicts those of us with filthy mitts on things and who think life *should* operate a certain way and to certain ratios. That birthdays are always happy. That a week in Thailand is meant to be relaxing. That a long-awaited date with your partner at a special restaurant will bring you closer together. When you shake things up there is no such expectation. It's so wrong it's right.

DO IT AT THE WRONG TIME

When I'm gripping too tight, I'll . . .

Eat dinner foods for breakfast. This morning I ate mashed pumpkin with garlic. Sometimes I eat grilled sardines on lentils. Or lamb chops.

Go to a 10am movie session on a sunny day.

Grocery shop at midnight. It feels slightly wrong and lonely but, equally, much more fun than going on Saturday morning.

Write beyond my 9pm screen curfew on a balmy night sitting outside with a glass of wine, because I'm in the moment. And then take the following morning off and go to yoga at "office o'clock."

SLEEP AT THE OTHER END OF THE BED

Doing a few things back-to-front helps, too. Like putting your head where your feet normally go when you climb into bed. This tiny change is like a holiday. Everything looks different in the morning. Fresh and lighter. I read recently that business and creativity coaches have caught on to this concept and are calling it "vu déjà." Which is the inverse of seeing something in the same way. It's seeing what you always see, but in a different way. Or in Zen parlance it's to have "fresh eyes."

space

104. I noticed some themes emerge when I asked people in the forums and roundtables I conducted what anxiety actually *feels* like. I also dug around to find what other writers on anxiety reported. It all revealed quite a lot.

First, the feedback was that it feels like a flood of thoughts:

> *Picture a bunch of people loudly talking to you about everything you don't want to hear—that's how it feels in my head.* – Anon

> *Thoughts flood and for me paranoia sets in and I try to grasp on to at least one thought I can be rational about.* – Pepperoni

> *[It's] like there are a hundred things needing my immediate attention and knowing that I can't attend to it all at once, including racing thoughts.* – Lisa Jane

> *Anxiety is like having new tabs opening very quickly [on your computer] one after another and not being able to close them or stop new ones from opening—but in your head.* – Anon

Then the flooding thoughts build, with nowhere to go:

> *Anxiety feels like being the passenger of a race car driver while pleading to be let out. I close my eyes and take deep breaths at every endless turn.* – Anon

> *For me it's like a boa constrictor around my body, getting tighter and tighter as more thoughts come into my head.* – Pepperoni

> *Everything, all of life, is crammed into a tube of toothpaste which has a caked-over nozzle.* – KT

Andrew Solomon, author of *The Noonday Demon: An Atlas of Depression* (and transfixing TED speaker), described it as "like wanting to vomit but not having a mouth."

I want to throw in here that the linguistic root of anxiety is the Greek word *angho*, which means "to squeeze." Interesting, hey.

Then the pressure causes the thoughts to get overcooked—mushy and all mixed in together:

> *I am poured out like water, and all my bones are out of joint: my heart is like wax; it is melted in the midst of my bowels.*
> – David, Psalm 22

> *A very tangled-up spiderweb and all the web is mixed up with lots of emotions and tangled all together. The more I try to untangle these webs I get caught up in another web.* – Sadgirl

MY BALL OF WOOL THEORY

For me, anxiety feels like a knotted ball of wool inside my being. The thoughts and commitments and competing considerations twist and entwine and "moosh" into a mess. The tiny fibers of each thought weave together and it all turns into a convoluted, frayed, knotted ball of wool—useless for knitting.

Recently, I learned Franz Kafka also had a ball of wool theory. For him, anxiety was "the feeling of having in the middle of my body a ball of wool that quickly winds itself up, its innumerable threads pulling from the surface of my body to itself."

It's impossible to know where all the knots start. Yet, we still try to find the original thread, somehow believing that once we find it, this one unifying explanation for everything, we can tug at it and have the ball unravel cleanly. We think the fix is linear like that. That one motivational philosophy or one successful relationship or one perfect job will straighten out the mess.

But I put it to you that messy balls of wool don't work like this. Nope. Our filthy-mitted meddling and tugging only tighten the knots more.

Instead, the only salve is to gently take the messy ball in both hands and tenderly loosen it, a bit at a time. The ball starts to unfurl and expand. It is still knotted, but not as tightly now.

After a while a whole section unfurls. And then another. Then, after much careful tending, one end of the string floats loose. Maybe the rest of the ball fully unfurls. Maybe it doesn't.

But the point is, the whole bloody knotted mess is looser now. There's more space.

105. If you're anxious, part of the healing journey is to create space. To soften and expand and back off from this drive to "fill" the space (in our guts, our diaries, our weekends, our wardrobes).

Studies have shown that particularly creative and anxious minds need a lot of space—or downtime—for what is called our Default Mode Network to make sense of things.

"Deprived of it we suffer a mental affliction as disfiguring as rickets," essayist Tim Kreider wrote in *The New York Times*. "[Space] is a necessary condition for standing back from life and seeing it whole, for making unexpected connections and waiting for the wild summer lightning strikes of inspiration."

Someone with bipolar once told me they need to be alone a lot so that they have space to play out the conversations in their head. They didn't say they needed time. I know it might sound like semantics, or seem metaphysical. It's not. It's an attitude. A feeling. Space implies gentle unfurling. Time speaks to pressure.

Most of us cry out for more *time,* thinking that's what we need (much like balance). But tell me when more time has helped anyone in a hot anxious mess? Time doesn't release the pressure. Time doesn't take the cap off the toothpaste. Time doesn't loosen the knots. If we get time, we tend to just fill it with more thoughts.

What we need is more space.

106. I used to see a high-strung Eastern medicine doctor during my Mid-thirties Meltdown. Dr. L would bustle in and yell at me that my spleen was stuck, "Very, very stuck!" and stick needles in my ovaries to "wake them up." Her pace was frenetic and high-pitched. I asked her how she did it. How she maintained her own health with such a frazzled energy (she also had Hashimoto's). Frazzle and Hashimoto's come in the same package. Anxiety is the special icing on top.

Dr. L told me she books out an hour every day to lie on the floor in her consult room to pause. She does nothing. She just lies there. Yes, it's "time" that she books out. But it becomes "space" when it's kept empty, as a vessel in which to simply stretch out a little. "Only fifteen minutes needed. Sometimes five," she yells at me. "Just the dark room is enough."

I remember also chatting to a straightlaced CEO mate of mine who has her PA book out fifteen minutes either side of every one of her appointments. "I use it to reflect on what just happened," she says. "It gives me the space to view what I need to do next."

I drove to visit family in Canberra recently. That's four hours of bleak nothingness that I'd normally fill with returning calls and listening to podcasts. This time I did nothing. Not even music. Just large expanses of sheep-ridden, dusty space. I didn't "use" the time. I just sat into the space. And fresh thoughts bubbled up from the nothingness.

And this is the beauty of space. It's a nothingness that surrounds and sits between all the somethingness in our lives. It might be a little stint in a dark room. A walk around the block between chapters. A quiet moment in the loo when the kids are watching *Play School*. A visit to an empty church on a lunchbreak. It's only in the nothingness that we can see the somethingness. Without space, it's like watching a movie three feet from the cinema screen. We can't see the whole picture. And we lose ourselves in the noise and the fuzzy pixilation.

When we have space we have a chance at having a better anxious journey.

FIND THE SPACE BETWEEN BREATHS

Try this. Just now. As you read this bit. Might as well. It builds on the previous breathing trick I shared earlier.

– Flick back to page 74.

Breathe in for a count of four.

At the top of your breath, *hold for a count of three.*

Gently, evenly, breathe out for four.

At the bottom, *hold for three.*

When I hold for three, I don't so much hold as pause gently in the space between breaths. I don't force the hold. I just suspend my breath. Try it.

Then let the energy swirl around in this space, filling it, expanding. Repeat for a minute or two with your eyes shut.

SMILE WITH YOUR EYES

Yes, try this one, too. It provides instant spaciousness.

Sit. Close your eyes. Now start to meditate in whichever way suits you. Or just sit and breathe in and out, doing the pause thing, as per above.

When the thoughts start to tumble in, and you start thinking of the tricky email you have to send later, and you give yourself a hard time for pretending not to see your annoying, but lonely, neighbor in the supermarket earlier . . . gently smile with your eyes.

Gently and softly.

Perhaps notice the way it releases the muscles in your jaw and at your brow. And the way it kind of makes your temples spread apart, making it feel like your mind is boundless, floating, with no skull trying to keep everything in. Notice the space in your skull. Notice the way it makes you see things with kindness and expansiveness and the way that little commentary voice that likes to judge things and have an opinion on everything goes back in its box for a bit.

In one famous 1993 experiment, researchers asked a bunch of people to smile forcibly for twenty seconds while tensing facial muscles around the eyes. They found that this simple, brief action stimulated brain activity associated with positive emotions. If you're still reading this, you might like to know that in the U.K. some scientists bothered to tally the mood-boosting value of receiving one smile. If the smile is from a friend, it's equal to the feel-good brain stimulation of 200 chocolate bars; if it comes from a baby it equates to 2,000 bars! There you go.

Inversely and just as randomly, a recent German study found that Botox injected into smile muscles interrupted the brain's happiness circuitry. Numbing our smile muscles made us sadder, more anxious. Which is a lovely irony: Aging gracefully really is the more joyful route.

And with that, let some sweet spaciousness pour over you. Just for a little moment.

boundary building

107. A makeup artist at the studio where I sometimes do morning television appearances tells me there is actually never any reason for her anxiety. "I'm not stressed, but I'm anxious." Yes, many of us just are. Today, anxiety is in our collective bones. It might not be clinical. We might not be able to point to the cause. We might be fatigued by the idea of "blaming" it on something, such as a shithouse childhood or emotionally limited parents. But it's there, everywhere. Everyone around me is talking about anxiety, as something that goes beyond their own personal struggle with it. It's a "thing." It's not a stretch to say that it's driving us all mental.

Shai, a very self-aware woman in her twenties who attended one of my forums, goes as far as saying her "everyday" anxiety, the kind she notes as a collective thing, is more distressing than her medical neuroses, for its pervasiveness: "I'm okay with dealing with my PTSD stuff . . . but it's the other kind of anxiety, the constant feeling of hypervigilance that's not linked to anything tangible or anxiety-inducing, that's hardest."

108. Yep, the world is getting more anxious. But it's not just something that's in the global waters. It's specifically a "Modern Life" thing, or more accurately, a Western Modern Life thing.

New findings from the most comprehensive meta-analysis of anxiety and depression research to date, published by researchers at the University of Queensland, show that clinical anxiety affects significantly larger percentages of the population in the U.S., Western Europe and Australia/New Zealand compared to the Middle East and Asia. Interestingly, the opposite was true for depression, with people in Western countries less likely to be depressed.

What gives?

Let's take a look at how Modern Life goes. Mostly, it's frenetic and at a pace that's not conducive to reflective thought.

> Working on the fly from laptops.
>
> Weaving in and out of traffic.
>
> Eating on the run.
>
> Walking around with takeaway coffees.
>
> Keeping up with technology updates (the Anxiety of Being Three Updates Out of Date!).
>
> Being expected to turn around school projects overnight (what's your problem? You have Google!).
>
> Ferrying children to violin lessons.
>
> Taking the whole family to Paris and London (for no reason other than everyone else seems to be doing it).
>
> Online grocery shopping on your lunch break.

It's all too fast for our human dimensions, as David Malouf put it. We don't have time to adjust, to work out our priorities, and to reflect on whether what we're doing when we're running around madly is actually meaningful to us.

While we are meant to have more time (all those time-saving devices were meant to deliver just this, no?), we have less space. We are "on" 24/7. Every gap is filled. Even waiting at bus stops. We don't leave work and unwind and stare into space for a bit, enjoying the sound of the birds, the soft dusk sunlight on fellow passenger's faces. Nope, we must prune our social feeds.

I know there's a generation of us who relish air travel because it's a rare chunk of time when we have a "gap"—we can't be contacted, the inbox influx abates and if we do take the opportunity to get through emails, we have time to edit them before they send. Lovely languid luxury, I tell you! But airlines are now phasing in Wi-Fi on planes.

Technology freed us up . . . to imprison us further. It's created the imperative to go faster, to take on more ideas, and to juggle more. There are no excuses for not coming up with an answer, and immediately. Not when there's Google. Or Siri. Or bots. There are no excuses for letting something slip. I mean, didn't it automatically upload in your online diary?

But what if we need more time to *know* and to *feel* if it's the right answer?

To stay on top of all the ideas and opportunities that Modern Life now affords us we have to keep multiple tabs open in our brains, which sees us toggle back and forth between tasks and commitments and thoughts.

And all of it competes. And it clusters. And down we go in a hyper-tabbed tangle, waking in the middle of the night so we can get through the backlog of thoughts and ideas and commitments.

All of which has made us too bloody overexcited. When we had tantrums as kids, Mom would say we were overexcited. "Come on, a little less excitement," she'd say. Yes, we need less excitement.

But self-mastery triumphs in this Modern Life of ours. So if we haven't found happiness or calm or balance amidst it all—if we don't cope—it's because we've not tried hard enough. Because Modern Life dictates there's an answer out there . . . you just have to try harder to find it and master it. Of course it doesn't exist. So we are set up to fail.

I feel for young people. I think they're hit particularly hard by this doomed imperative. Many sociologists peg increased anxiety among teens and young adults to this phenomenon.

The standard solution is to consume—food, possessions, partners, gurus. If our self-worth is suffering, we're told to buy a new moisturizer. Mark Manson, author of *The Subtle Art of Not Giving a F*ck*, writes, "We have so much fucking stuff and so many opportunities that we don't even know what to give a fuck about anymore."

Shai once again: "Today we're told to do more stuff that has no purpose, which makes us anxious."

Again, I think young people feel this acutely.

And here's the dirty clincher: All of it drives us outward, away from our true selves and from our yearning to know ourselves better. Plus, it drives us away from each other. Lack of community and belonging is cited by Dr. Jean Twenge, a social psychologist at San Diego State University and author of *Generation Me: Why Today's Young Americans Are More Confident, Assertive, Entitled—and More Miserable Than Ever Before*, as the primary driver of anxiety today. I'd include extensive quotes from Dr. Twenge, but I think the book title says it all.

Then (big sigh), when we do find it all too much, Modern Life slaps us with a "disorder" or disease diagnosis.

When really, it is quite fair enough—reasonable and sane, even—to find ourselves anxious when faced with Modern Life. I mean, we're too bloody overexcited. Right Mom? And as many pragmatic mothers tend to say, "Now what are you going to do about it?"

109. There's a lovely Ayurvedic way of looking at the whole anxious caper that in recent years I've found to be extremely helpful.

According to Ayurveda, anxiety is not a disease. It's not an unhealable disorder. It's merely a symptom of having got a bit off balance. We don't fix anxiety. It doesn't need a fix. It just requires a bit of rebalancing.

In the Ayurvedic tradition, we all work to three doshas, or "types": vata, pitta and kapha. This is less woo-woo than it sounds, I promise. It's simply a way to categorize body/personality types that exist for a multitude of evolutionary reasons. We all possess a mix of all three doshas, but tend to have one that dominates. Our dominant dosha can get out of balance, which causes us different digestion/weight, health and emotional issues.

Make sense?

So, generally . . .

Vata types: have light, flexible bodies and big, protruding teeth; small, recessed, dry eyes; irregular appetite and thirst; often experience digestive and malabsorption problems; are easily excited; alert and quick to act without much thinking; may give a wrong answer but with great confidence. Their dominant force is wind so they do not like sitting idle, and seek constant action. They're FLIGHTY! Vatas hate cold. Hate, hate, hate it. They need warm, mushy foods to bring them back down to Earth. And they love summer.

Pitta types: have a medium frame and weight. They seldom gain or lose much weight. Their eyes are bright but tend to be sensitive to light. Pitta people usually have a strong appetite and thirst. They have excellent abilities for learning, understanding and concentrating; are highly disciplined; can be judgmental, critical and perfectionist, and tend to become ANGRY easily; have moderate strength, and a medium lifespan. Their force is fire—so summer is the time when pittas get easily aggravated. Sunburn, poison ivy, prickly heat and short tempers are common. Pittas need and love cooling foods (salads), and should avoid chillies and hot spices. I've noticed pitta men are often bald . . . too much heat coming out the tops of their heads!

Kapha types: have a strong and large body frame, big eyes, strong teeth and thick, curly hair; thick, smooth, oily and hairy skin; slow digestion and metabolism, which often results in weight gain; cravings for sweet and salt; a calm, steady mind; a deep melodious voice and a monotonous pattern of speech. Kaphas are an earthy type and can get heavy—they need firing up. They respond well to coffee and spices. Kaphas tend to get aggravated as the moon gets full and during the winter and early spring, when the weather is heavy, wet and cloudy . . . it makes them too heavy and damp.

But here's the thing. Excitable vata controls all the doshas. Like oxygen, it's vital to everything else.

And here's the other thing. When vata is out of whack—when it gets overexcited—we get anxious. No matter our dosha. Vata is the oxygen to the fire (pitta), the wind to the earth (kapha). Too much and off we burn, or whip into a whirly-whirly, out of control.

And here's one more thing. Modern Life aggravates vata:

Abrasive city noises

Overstimulating coffee and sugar

Frenetic emailing

Cold gushes of air conditioning

Cold, crunchy dry foods (think chips, crackers, salads)

Multitasking and toggling between tabs

The worst is movement. Crazy cab rides across the city, overnight interstate trips, flying, driving long distances—they all send vata over the edge.

So add up all the things and what do you get? A whole bunch of us suffering from vata-deranged anxiety.

When this first occurred to me, I felt disappointment. I'm good at moving fast, it's what defines me. I like the agility of being a windy type. As I've mentioned a few times, I'm arrogantly attached to many of the factors that make me anxious—the speed, the multitasking, the constant change. Which might sound ridiculous to the nonanxious. Not so much to those of us who rely on these things as coping mechanisms.

But I soon learned that Vedic philosophy doesn't ask you to retreat to a cave or to renounce your mobile phone. It just suggests that when you're out of whack you tame your vata a little with a bunch of techniques, to steer things back to a bit more balance.

HOW TO TAME YOUR VATA

This is how I do it. It's cumulative. The more you do, the more your flappy vata kite will chill. And the more balanced you'll be. And the better you will feel. And the more your anxiety will balance out . . .

I avoid air-con and fans. And get out of the wind. I often carry a shawl or spare sweater to wrap around my neck when I'm anxious in case I find myself imprisoned in a draft against my will. I'm aware that spring is not great for me. The wind is likely to send me off-kilter. This is fine if I'm aware of it. I just do more of the stuff on this little checklist.

I back off from coffee when I'm fretty. I often get asked about coffee. Is it bad? Should you drink it? I take my cues from Ayurveda. Nothing is inherently bad; it's about whether it's sending you off the air. If you're asking if it's bad, it might mean that you feel that it quite possibly is. Me, I drink a long black most days. But I intuitively back off from it on a regular basis, in part to ensure I don't get addicted to it, and also to rest my adrenals and let the wind out of the kite for a bit.

I try to eat lunch at 1pm. And dinner at 7pm. You pick your own timings. The routine bit is key. Vata energy needs routine to be calmed.

I eat heavier foods—sweet potato mash, root vegetable soups thickened with smooth yogurt, and porridge. They get vata energy heavy and grounded—veritable Miss Janes of the food realm.

I eat oil. I eat coconut oil and ghee in stews and curries. And pour olive oil on my vegetables every night. A tablespoon or so. But when I'm worked up, I increase the amount. Oil nourishes and warms us. In Western nutrition, this is also acknowledged. Many of the most essential vitamins and minerals, and proteins, are fat-soluble only. Which is why the Greeks and Italians always add oil to their veggies, the French and our grandmothers added butter. We need fat to be densely nourished.

I sit still for 5–20 minutes several times a day. I meditate. You might want to just sit on a small bench with yourself. Or lie in a dark room.

I tell friends I have to leave by 9pm when I'm out at night. Vata is best pacified into sleep during the "kapha" period before 10pm.

I turn off social media on the weekend and after 8pm at night. Sometimes, when vata (anxiety) has escalated, I commit Safari suicide, or Google Armageddon, which is to say I shut my browser down and let all the tabs go down with it. Everything is findable again. I let it go.

I wear vests and things that my mother always told me would keep my kidneys warm. I also bought a pair of Ugg boots three winters ago that extend to the knees.

I walk everywhere I can. I avoid car travel, even bike travel when worked up. In Ayurveda, the belief is that anything faster than walking can throw us off.

I do yoga. Preferably hot to really encourage getting down low and heavy.

I don't go to The Shops. Shops are dens of vata and anxiety-inducing pain. There's the air-con, the noise, the bright lights, the options.

110. Blaise Pascal was a French philosopher who was plagued with illness and anxiety and jumped between an esoteric collection of obsessions. He studied "the triangle," invented mechanical calculators, and wrote extensive spiritual polemics, among many, many other unrelated pursuits. He famously wrote, clearly from experience:

> Distraction is the only thing that consoles us for our miseries and yet it is itself the greatest of our miseries.

— **cruel irony #16**

The nervous toggler Pascal also remarked that all of man's problems come from his inability to sit quietly in a room alone. And to let nothing happen. I'm not sure he ever got around to realizing such a skill. He died at thirty-nine from what has been interpreted by some as anxiety-related complications.

To this day, the idea of letting nothing happen challenges me. When I was a little girl living in the bush I would jump with excitement when the phone rang and physically ached to hear the sound of a car rumbling up our long driveway. I would climb a tree and wait and listen. For hours on end. For something to happen. *Someone's coming! Something's about to happen!* We rarely had visitors, perhaps once every two or three months. I'd cook oversalted scones and soggy cakes for them and preplan my outfits and conversations with the adults. I don't think this anxious, incomplete anticipation has ever left me.

Today, when I'm jumpy, I grab my phone to see if something has happened on social media. Has someone responded to my Facebook post? Has my latest Instagram Stories share been flooded with views? I've witnessed my reaction sometimes to returning to my car when I haven't put money in the parking meter. I actually *like* the feeling of the impending drama of a ticket! I can only conclude that it's an ingrained addiction to "something about to happen."

I met Oprah's life coach once. Martha Beck writes a column each month in Oprah's *O Magazine,* which I was obsessed with at the time. She was the most heart-forward, funny writer I'd come across. She's a thin woman with soft red hair and green eyes who reminds me of a little bird. She darts about and speaks fast, but somehow delicately. It turns out she showed me she can bend a spoon using her mind (and her tiny hands, lightly). I'm not kidding. A few things you should know: She first gets me to try bending the spoon (there was no way I can budge it, and, let me tell you, I have "man hands"); Martha weighs little more than a whippet with the corporeal strength to match; the act is un-premeditated, using a spoon from the café (not her own); and, finally, if you've seen *The Matrix,* yes, it's rather like that.

Anyway. That's not the point of this tale.

I'd asked if we could meet up during one of my Manhattan visits. It was during a bit of a transition period out of my Mid-thirties Meltdown. I was gathering strength. I'd started meditating. I was writing my "life-bettering" column and was in New York to interview a few Big Name Life Betterers. When Martha and I meet in the Hudson Hotel for green tea the first thing she says to me is, "I could tell you were one of my team." She reckons she knew from the way I'd phrased my email. She described me as "leaning so far forward it's not right." She got me on that one. Even before meeting me she could tell that I was hypervigilant and overtoggled and leaned into life with a ferocity that frightens most people and that I try to tame, but also know is a beast worthy of being called beautiful at times. It reflected on her maturity and wisdom that she acknowledged she was in the same camp. She was reaching out, as if to say, in the kindest way possible: "I know . . ."

Once again inspired by — Alain de Botton.

I don't think it's bad to lean forward, or to enthusiastically make soggy cakes for house guests, or to want to know what everyone else is doing with their Sunday afternoon while you're scrubbing the bathroom. We're human. We're curious and we reach out. It no longer serves us, however, when we do it to run from something. And if we're too overexcited and wound up to know that we're doing it. And if we've wound ourselves up so much that we're no longer able to sit with ourselves and let nothing happen. Only we can tell if that line has been crossed. As we start to sit and stay and settle and find space and unfurl and tilt and reflect discerningly, we're able to feel when we have. It hits us as an *ugghhhh*. We've lost ourselves again!

To be honest, even meditation can be a form of reaching outward. British mindfulness counselor Richard Gilpin writes in *Mindfulness for Unravelling Anxiety* that it's a common mistake among those wanting to get mindful with their angst to expect to achieve states of calm through meditation. "This is a form of grasping—a seeking to indulge in pleasant states and to avoid the unpleasant," he writes. Meditation retreats around the Eastern hotspots are full of people running from something, hiding behind Magic Happens bumper stickers and quasi-spiritual hash tags (#blessed #lightwarrior #unicornsandrainbows).

"A wiser orientation would be to appreciate (and investigate) calm states when they do arise and to treat anxious ones with great kindness and respect. The radical encouragement of the practice is to sit with the most disagreeable of states for as long as they last. Sooner or later, they exhaust themselves of energy."

READ THIS. IT'S FUN.

It's an exercise I learned from Gilpin:

When you get to the end of this sentence, focus your eyes on the full stop after the word "up," stay right there, don't do anything else, and watch what shows up.

Now that you are reading this sentence, did you catch the moment when your eyes moved from the full stop? Was that movement intentional or did it seem to happen by itself?

Now cast your mind back to when you were focusing on the full stop. What happened? What did you experience?

Would you like to try the exercise again? Please do. It will be different this time. I don't want to say how or why, because that's not the point. I'd be allowing you an opportunity to grasp out to someone else if I did.

111. Here's what I reckon. In the olden days, there were institutional boundaries that kept us from getting overexcited. Our employers provided them, for instance. We worked a nine-to-five day and weekends were off-limits. We didn't take our phones home. We didn't have home computers. We weren't on call 24/7. We could whine if the boundaries were crossed and someone—a boss, the HR department or the Union—would fix the issue.

The Church ordained days of rest each week and the shops were closed on Sundays. Plane trips were once-in-a-lifetime experiences. We communicated with letters, written slowly and mindfully. No one expected one-hour-or-less response times.

These boundaries created certainty anchors and reduced the number of decisions we had to make. They helped us keep on an even keel. But today there are few such boundaries. Information and obligations flood in. We keep thinking, after all these years (decades?) that there will come a lovely fine day when the influx eases. When we'll get on top of it, as we once used to be able to. We still work to this old-school notion. But such a day no longer exists.

I think we're also working under the misapprehension that someone else will come along soon and create the boundaries for us, like in the olden days. And police them. I have news for you: they won't.

What we're yet to work out is that we have to create the boundaries ourselves. This is the new barometer of success, wellness and happiness: How well can you create your own ways to shut down the distractions, reduce the toggling, stem the tide of frazzling data, carve out space in your week for reflection and stillness?

In the past, success was gauged by how well you could *hunt down* information, collate data, find a great reference in the *World Book Encyclopedia*.

Now, success must be gauged by how much information and data you can *shut out* . . . via your own boundaries. When you realize this, it's actually quite freeing. We're given permission! You mean I don't have to wait for anyone else to fix this? I actually *have* to do it myself. You beaut!

BUILD YOUR OWN BOUNDARIES

You have to create the spacious, loose, still, tabless world you want to live in. You just have to. By way of inspiration, I'll share with you that the most successful people I know have created firm boundaries for themselves. Very firm. Here's a few examples that I found particularly helpful. You might, too. Pick the ones that appeal.

Check your emails twice a day only. The 4-Hour Workweek's Tim Ferriss told me about this one. He checks his in batches at 10am and 4pm. He has an out-of-office on permanently that advises anyone who contacts him of this, so as to manage their expectations. When we chatted over the phone a few years back (after he replied to me thirty-six seconds after I sent my email, at 2:45pm San Francisco time) he told me that creating boundaries for himself meant training those around him to not expect him to be always on.

Try the 10am Rule. One self-styled life expert I came across says to not "react to anything until 10am." That is, first do the stuff that matters to you, rather than knee-jerking out of the gates to the demands of others.

Live somewhere slow. Zen blogger Leo Babauta moved to Hawaii to get away from the distraction that is city life. Novelist Pico Iyer moved from Manhattan to rural Japan, "in part so I could more easily survive for long stretches entirely on foot, and every trip to the movies would be an event . . . Nothing makes me feel better—calmer, clearer and happier—than being in one place."

I moved to the beaches north of Sydney so I could have less city frazzle around me and ready access to the healing effects of the bush and the ocean. It was a big sacrifice. It meant leaving behind friends and my local community and living temporarily again. I did this just as I embarked on researching and writing this book and I'm conscious this boundary has had a big part to play in providing *just* enough grounding to get me through the anxiety of writing a book on anxiety.

Have a Family Investment Bucket. My mates Nicho and Heidi initiated this in their house. Phones and iPads go in there from 6pm until the next morning.

Leave your phone at home. When I'm feeling wobbly I leave my phone at home. It's a defiant act in this day and age. Every week or so I drop out completely and don't answer my phone all day.

Get a room of your own. I met someone who, every few months, books in to a cheap hotel not far from her home on wotif.com. She checks in after work. Runs a bath. Reads. Paints her nails. Orders room service. Watches a movie. Sleeps soundly. And checks out the next day and goes to the office. She tells her husband she's away for work for the night. She feels a little guilty about lying to her family. But feels that the payoff for all involved is worth it.

Try a Think Week. Microsoft's Bill Gates has one every six months. He extracts himself from his chino-wearing Silicon Valley brethren and heads to a wee cabin on a hill, eliminating all distractions. Lifehacker blogger Michael Karnjanaprakorn describes his own attempt at such a week: "I created a life to-do list, did a lot of research on happiness (where I learned that it's about the frequency, not the quality of positive experiences). I focused on my personal development (not career development). I went for a hike in the woods. I learned how to cook organic food. I read three books I've been meaning to finish forever. I did yoga and meditation every day, which cleared my mind. And I sat for hours and just stared at my beautiful surroundings during the morning sunrise." I've done this a few times. It's challenging.

Create your own Sabbath. I joined *National Geographic* explorer Dan Buettner on one of his trips to what he terms Blue Zones—pockets of the planet where people live the longest. He drummed into me that one of the most important of the nine factors that lead to a long, healthy life is having a day of rest. Me, I take Thursdays off. It's marked out in my calendar as "busy." Permanently. On Thursdays I catch up on long reads I've saved. I'll nut out thoughts in a notebook. I'll do things back to front. I wrote this book on Thursdays. Thursdays are a day of space.

Create a mercenary Out-of-Office notification. When I travel or am in writing shutdown, I set an Out-of-Office reply advising that I will not be following up on emails. I invite people to resend their request after my return date if the matter is still important. Which effectively pushes the onus back on the sender to reconnect. Which is smart. I mean, how do I know if they've resolved the matter in my absence? And, after all, they are the ones needing a piece of me. We forget this. We forget that email is not a summons.

Don't be Google. People can get lazy with emails, firing off a question that can easily be Googled or nutted out with a little time and care. Anxious people tend to be overly earnest in their desire to reply and to help others. Create a boundary around this. I do. I used to reply politely to these emails. Now I just press delete. A journalist friend, James, says he feels justified in being an email nonreplier. He explained it well: "People think that because they've spent five seconds firing off an email asking something of me, they deserve a response that will take me twenty-five minutes to research and compose," he says. "It doesn't weigh up."

Just write less emails. Every email sent begets another three, filling up your inbox. See how it feels to pull back a little. Observe how many issues sort themselves out without your vigilance.

Own less. I've spoken to many digital nomads and minimalists who've declared light living is the way to go. They banged out (sponsored) blogs and got their exploratory trips around the world paid for. Maybe not surprisingly, most of them gave up the experiment after a year or so. That said, I personally believe wholeheartedly that living simply is the way to live. For me, it reduces a lot of decisions (I don't own enough clothes to have "wardrobe crises" when I'm getting dressed) and steers us to living to better principles. It creates space to do so. After the global financial crisis, a host of studies came out showing that we are indeed happier with fewer possessions. Most showed it's due to "hedonic adaptation"—we're programmed to stabilize happiness levels. So the happy jolt from buying or having stuff is always short-lived. Which is why a study published in the *Journal of Consumer Psychology* found that experiences (a canoe trip) make us far happier than things. Further, another study published in *Psychological Science* found that more possessions—"an embarrassment of riches"—reduces our ability to enjoy simple things, like sunsets and chocolate. The awareness of having stuff distracts us from basic pleasure.

Many experts in the realm, however, bang on about the joys of decluttering and tidying up. I will pipe up here and say that chucking stuff out and shedding our shackles is not the responsible way to go about it. Discarding good resources and getting obsessed with reordering things into boxes (which you go buy from container shops) is plain wrong. Don't buy the crap in the first place is my antidote. Or more directly, don't go to the shops.

112. There's a farmer on his deathbed. He's gasping his last breaths, but has assembled his greedy, lazy sons to tell them where he's hidden the family fortune. They hover impatiently, expectantly. The farmer raises his arm and points out the window to a large field. "It's out there." And with that he passes away. The greedy, lazy sons immediately begin digging up the field, row by row, searching for the treasure. Finally, they get to the last row and are bitterly disappointed. The treasure wasn't found.

The following harvest they have their best crop ever.

Do the journey. Do the work. Do the little right moves. The crop comes.

the wobbliest table at the café

113. My word, I'm irritable. Not just now. Often. Many of my anxious friends are, too.

Heavy breathers in yoga kill me. So do *hmph*-ers and nose whistlers.

When dinner companions do that thing they do in American movies where they shovel food in their gobs as they talk and wave their fork wildly, their head flung back, making schmucking noises with their gums . . . yeah, that, *THAT* sends me mad.

Don't get me started on foot tapping, leg jigging, and finger drumming on communal tables. I'm not alone. I put these afflictions into Google along with "anxiety" and it spits back a raft of forums, studies and blog posts on the matter. I learned that anxiety widens personal space—we need more than the standard 8–16 inches that the average person requires to feel comfortable. I also learned that those of us who veer into mania and hypomania generally find most aspects of sharing the planet with others irritating because, as Jay Griffiths writes in her manic depression memoir *Tristimania* (tristimania is an eighteenth-century term for bipolar), "when you're racing and overcapable and wildly energetic, any ordinary human speed looks like lethargy and . . . feeds the irritability."

Fans and breezes send my peripheral nerve endings into an agitated spin.

Stereo speakers that pump the bass in such a way that a tangible vibration overlays the whole experience wholly distract me.

As do all vibrations, audible or tangible. Air-conditioning units at night will reduce me to tears. I hear it and feel it in my viscera. Industrial-strength earplugs can't protect me. And, as I've mentioned already, even just someone's heartbeat felt through the mattress when they lie next to me at night can leave me awake all night.

Anxiety, as we know, activates the stress response, which immediately causes a heightening of our senses and stimulates the nervous system so we are keenly aware of, and have enhanced ability to defend ourselves against, danger. So all of the above makes biological sense. This is vaguely comforting.

I don't know about you, but my irritation is amplified when I'm anxious. These sensory triggers, however, can also send me into anxiety. And around and around we go. I often feel like a bundle of nerves that's been scrubbed with steel wool until thoroughly raw and exposed. And that to step out into the day is to be dipped in a drum of acid.

My friend Kerry said this to me once: "It's not stress that makes you stressed. The experience of being human is what makes you stressed." She told me this over the phone as I sat in a share car in a service station car park, paralyzed in one of my anxious spirals with the car due back in its pod fifteen minutes ago. Kerry's someone who can pull back and see the forest when I'm tangled in the trees.

I can't recall what humanoid-y reality had tipped me on this particular occasion. No doubt it was something inversely trifling compared to the panic it had caused. But I remember her words hit a special spot that afternoon.

"That's exactly it," I said breathlessly.

Some of the most anxious people I know do public speaking for a living. Others are actors on big stages, or balance three businesses or compete in international sporting events. But a dripping tap late at night or a crease in their sheets or the sound of a work colleague's ring tone will send them over the edge.

And then there are smells. Artificial fragrances—in perfume, washing powders on hotel sheets, in the shampoo in the hair of the person swimming past me in the ocean pool and in those scented sticks people have in their toilets—leave me physically ill. I'll get hugged by someone who sprayed perfume on themselves five hours earlier and the smell will leave me dizzy and agitated until I wash it off. I stand back when people go in for the hug. And when I'm anxious I can't go to dinner or movie theaters with perfumed friends.

I've looked into why this happens, and why smell can so readily trigger anxiety, for me and many others. Apparently, when we emerged from the primordial soup our gnarly old amygdala evolved from our olfactory bulb and both now sit in the deep core of our noggins. Anxiety, then, can see our emotional system get intertwined with the olfactory processing system. So smells easily—and instantly—are associated with certain fears. University of Wisconsin-Madison research shows that the olfactory bulb also has direct access to the hippocampus, which is responsible for associative learning, which explains why a smell can then trigger anxiety. Yep, around and around we go.

But I wonder how much of our tedious sensory "specialness" is a legitimate intolerance to the 800,000-odd unregulated toxic smells that we're smacked in the face with. We anxious folk are the canaries down the mine, performing a community duty, perhaps. Our hypersensitivity once warned others in the community against being poisoned by dangerous mushrooms and attacks in the night by stampeding rhinos. It does much the same today, tirelessly flagging modern toxins that are making us unwell.

Performing this duty, however, makes me feel like a prisoner in my own body. I'm often forced to isolate myself when my sensitivity flares. Not just because the world hurts too much. I'm also self-conscious of how precious I seem to those around me. And so I spare everyone the pain of it all by disappearing for a bit.

Know what I mean?

114. **Monachopsis:** *(noun)* The subtle but persistent feeling of being out of place, as maladapted to your surroundings as a seal on a beach—lumbering, clumsy, easily distracted, huddled in the company of other misfits, unable to recognize the ambient roar of your intended habitat, in which you'd be fluidly, brilliantly, effortlessly at home.

— The Dictionary of Obscure Sorrows

115. How to make peace with all this? I've toyed with this conundrum. I have to live on this planet with other humans; I *want* to live on this planet with other humans! And I can't keep running. Wherever I go, there I am. So, too, the humming air-con units. And the other humans.

Let's go back to my deeply uncomfortable month in the ashram in India. On my third day I asked to change rooms. The mold and the mosquitoes had got to me. I was convinced the room I'd been allocated was in the swampiest corner of the clinic. The hum of the generator two fields away that no one but me could hear kept me awake, even through earplugs. It tormented like a mosquito in a tent. In my new room, the insects were just as bad and the winds carried the rattling generator hum in a beeline to my room. Oh, and the old man in the room next to me snored every night.

Dr. Ramadas visited me on my fifth day. I was in tears, exhausted, feeling trapped. Why do I get into this position? Why don't other people get like this? Why can they sleep through humming? What will rescue me from this torture? "Our dear Sarah, stop asking why," he said. "You have learned all the knowledge, you have enough knowledge. And no answer comes, yes? You have to sit."

I had run 6,000 miles from my apartment in Sydney where I hadn't slept in years (the heavy-footed neighbors upstairs and the rattling hum of the navy ships a block away had tortured me). Before that I'd run from another house where the clang of the water pipes kept me alert all night.

He told me that I could move again. Or I could even fly home and give up on the treatment. But he pointed out the obvious. I'd run out of places to run to. "You keep moving. But it hasn't worked for you. The irritation has just followed you. The problem has to be healed and can only be done when it's in front of you."

He was telling me I had to stay. For a while there, I'd forgotten the importance of this. As we all tend to, I think.

So I stayed. I sat in the grimness.

116. I acknowledge this idea—of sitting for ages in grimness—is not overly appealing to most. Simply reverting back to our Facebook feed or heading to the mall to buy a moisturizer seems a far sexier fix.

Previously, despite all my *talking* about dealing with my fears and anxieties, I'd eventually run. I sat in the Hare Krishna camp. I sat with my Hashimoto's diagnosis. But only for so long. Then I'd ricochet off again. This time, in India, I ran the full experiment.

In my dank little room, staring at the ceiling fan, there was nothing to check, nothing that was going to happen, nothing to distract myself with. I had to pass through the waves of anxiety, one after another. I jerked away from it every 8.5 seconds or so, wanting to fight it, ask more questions and tell someone it was unfair. I watched the tussle going on inside me. But there was nowhere to run and no one to hear me. I stayed and I stayed and I stayed. I didn't like it one bit. My head got itchier, the drone louder, the thoughts faster.

But a few things were different this time. First, I was clearly rendered choiceless. Or, rather, I'd rendered myself choiceless (by choosing the strictest clinic I could find on the internet in the most unglamorous locale on Earth). Second, I kept on saying to myself, "Let's just see what happens." Third, I was strangely motivated by being around others doing the same experiment. We'd all signed up and committed, and in this context I realized that sitting in grimness made profound sense. Of course we have to go through the struggle. In the context of Modern Life, fueled as it is by distractions, such a notion is ludicrous, unpalatable, seemingly ineffective.

I heard about one woman who didn't emerge, not even for group meditation, for three months. She barely even got off her bed. I was inspired and stayed an additional week.

Finally, in my last week at the clinic, the sun came out and I was allowed to sit outside in the quadrangle on a plastic chair, with a rag wrapped around my head so the breeze didn't aggravate my vata.

I was still by now. I was also not loving it. I was both. The irritations were still there, but I was coping. The anxiety was far duller. It would appear my experiment was paying off.

I stared into nothingness, sitting there in the warmth. And then, I kid you not, I saw a dog turd rolling *up* the path. I looked closer. A beetle not much bigger than a ladybug was pushing the dried shit uphill. With his hind legs. I watched it for an hour, laughing on my own, resisting the urge to pick the damn thing up and put it at the top of the path for the poor beetle. That would have ruined the perfect metaphor.

117. During that time when my autoimmune disease slammed me to a halt and I was struggling to sit in the grimness of it all, I went to a close friend's three-year-old's birthday party. Mercifully there was no hired fairy or Miley Cyrus impersonator to entertain the kids. It was an organic, simple affair. In one corner of the park one of the toddlers had found a boggy patch of lawn—the only boggy patch of lawn in the park—and had plonked herself down in the mud. The sludge was oozing up through the sides of her nappy. But she was oblivious to any discomfort, happily playing with two sticks, engaged in some imaginary stick dialogue. Actually, to me and the two dads standing with me watching her, she seemed to be happy *in spite of* the mud. Actually, she'd elevated the situation further than that for herself—she was happy *because of* the mud. Like all kids, with much of what they do, they have a knack, or wisdom, for pushing through the annoyance (a fly in their eye, snot running down their face, a too-high step) to the happiness. It's the stuff we adults marvel at.

I remember reflecting that sitting in discomfort isn't just about lessening its impact through exposure. It can also bring about a very particular joy.

118. Which reminded me of the time I fell asleep on an ants' nest in the dirt on the edge of a cliff.

Back in my magazine days, I used to do this thing every other weekend or so, mostly in a gallant, agitated attempt to buck me out of an anxious rut. I'd head off on the train with my mountain bike to ride through the Blue Mountains just northwest of Sydney. I would eat a muffin on the train and drink a tragically burnt coffee from one of the kiosks at Central Station sitting in a seat in the sun. And I'd read the Saturday papers. I wore an old pair of bike shorts that were stretched out of shape and my hair plaited. I hopped off at stations along the way and rode for hours back down trails through forests and creek beds, along cliff edges, out of mobile range, alone. It all beat my anxiety into the background.

I'd ride for hours like this, train stop to train stop, jumping on the train back up the hill to start another ride before heading to a pub at the top of the mountain range. My bed dipped in the middle and—gloriously—had a chenille bedspread with a Country Life soap in a packet placed on a washcloth at the foot. I had a pair of clean undies, a T-shirt and a toothbrush stuffed in my camel pack. In front of the open fire I ordered nachos, and the steak (always overcooked) and a glass of red cask wine. And the apple crumble.

On this hot early autumn morning I'd been riding a thrilling single trail for two hours and emerged into a sunny clearing and it grabbed me that I should stop right there and lie down for a bit. My head rested against a log at a jaw-locking angle and my legs stretched out over the rocky soil. My body sank. Ants crawled up the inside of my legs and sweat was starting to congeal with the dirt in my hair and run in muddy rivulets down my neck.

It was imperfect. Messy. Wrong. And wonderfully so.

A whip-bird cracked far off in the cool of the tea trees. But there on my log in the hot sun everything was still. I looked up at the sky. Rocks were poking through the wad of padding in my shorts. I was happy. Quenched. My anxious buzz from a week of magazine-world frustrations and frenzy backed off. And I snoozed in happy delirium for fifteen minutes.

When we choose to go grim and lo-fi like this we lower our usual expectations, so that simple joys—sunlight, the stillness, the glow of the open fire, a turd being pushed uphill—become wonderfully apparent. With lower expectations there's less imperative to make things perfect. We can release our grip. We are in life, in its flow. We're sitting with ourselves. We let out a sigh.

119. I also have a "wobbliest table in the café" theory. It helps explain how we can apply this notion of sitting in discomfort to our anxiety on an everyday basis. We don't all have opportunities to be locked down in an ashram. We have to work with where we are.

I developed this theory when I finally got fed up with the horribly bourgeois perfect-café-experience indecision dilemma I mentioned several chapters back. I cringe to raise it again with you.

However . . .

After months of stalling outside cafés, unable to decide if it was the right one for me, I. Finally. Just. Walk. In. To. A. Café. It was a cafe I'd never noticed before. I choose it for this reason. Commercial FM plays on the radio—Enya's "Orinoco Flow"—and there are five flavors of focaccia on the menu all involving some form of jarred vinegary antipasto. I stride over to the first table. It wobbles. It wobbles from the stem and can't even be fixed with a folded-up bit of the racing section of the paper under the foot. The speaker above me crackles as Enya hits her crescendo. I don't move. I wait to see what happens. It's an experiment. I can do this.

The tea arrives lukewarm, but because this is an experiment it doesn't matter so much. I smile and the pubescent kid in his grotty apron smiles at me and asks if I'd like to try one of the savory muffins that broke coming out of the oven. You know, for free. And I say I'd love that and it arrives hot and buttery. And, in that very instant, I'm overwhelmed with joy.

You can spend a lot of energy avoiding wobbly tables. And you can fuss about with folded up bits of newspaper. But then, once the table is stabilized, you notice that smoke from the smokers at the next table is blowing right into your face. So you switch seats. Now you're in the path of the gale-force fan blowing in the corner. And your toast arrives burnt. And on and on it can go.

Or you can go straight to grim and lo-fi. That is, straight to what makes you anxious—in my case, choices, uncertainty, finding perfect moments and sensory irritations. When I sat at that wobbly table my irritation mounted as the table wobbled each time I put down my teacup. But I sat longer. It felt crappy and wrong and my body got prickly with anxiety as I sat there. But I sat longer in the grimness. You might try tackling an imperfect sleeping arrangement (noisy, no blinds) in this way. You might drop into a party, last minute (lots of people you don't know). In psychology circles this kind of experimenting is called "distress tolerance" and entails working with your specialist to remain in anxiety-provoking situations until your fear capacity becomes exhausted. Which it does.

The problem is that if you're anxious, you tend to flee (or fight or freeze) before you give the distress tolerance mechanism time to play out. I find this an enthralling idea. I mean, what if our inability to deal with our triggers came down to the simple fact we're unable to sit long enough? Actually, that's (pretty much exactly) what I'm trying to say here.

Long-standing, highly reactive or gnarly triggers are best dealt with alongside a counselor or doctor. But you can work the experiment on a few less intense anxiety-provoking scenarios and see how it goes.

You keep it casual, with few expectations, so you don't have to extend yourself too far. But the point is to actively *seek out* the discomfort so that you can *choose* to sit in it and do the experiment. Because you've chosen to do it, you're that bit more empowered. Also remember, it's just an experiment, to see what happens. Nothing more. You're just going to see what happens.

If you have claustrophobia, you can practice distress tolerance on a long-haul flight. You can do a bunch of things that first distract you (movies, podcasts, wearing an eye mask so you don't see the perceived small space), then soothe with sensations that effectively tell your prefrontal cortex that there is no emergency to get worked up over. If you are sipping hot tea under a soft blanket, then there must be no reason to run at full speed to the nearest cave!

For me, the fact it's a little experiment makes the grimness and the frustration of resisting my need to grasp and fix things a little more bearable. My metamission is simply to stay. And see what happens. So the quality of the tea, the comfort of my perch and the wobbliness of the table almost doesn't matter. I back the fuck off.

Back at the focaccia café that morning, I sat longer. And longer. By now twenty minutes had passed. And I felt the feedback loop that connects my anxiety to fleeing and fixing and grasping weaken with every additional minute that I stayed. I'm serious, I felt a distinct release inside my brain. It's all really crap, but I'm coping. And you know what this does? It gives me the confidence to settle even further. I get a jolt of satisfaction from this.

The pressure releases, the potency lessens. It doesn't matter. None of it matters. And if it all can matter less, the anxiety abates.

I remember as I ate my buttery muffin that the decades of gripping at perfection I was so used to seemed, well, boring. And kind of comical. And when something is a bit boring and kind of comical, it's no longer very potent. It's the same with recovering from a breakup. One day your clinging to your ex becomes—suddenly, overwhelmingly—boring. You hit saturation point on your obsessing and whining to friends. In fact, you realize that you were also bored in the relationship and so you see the funny side of being so attached to someone who was clearly not at your level. Phew. Of course, you only get to this point by sitting in the grim. Time, as they say, is the only cure for a broken heart.

Sitting in grim is also a defiant two-fingered up yours to your anxiety. I think this is great. For an added bonus, the practice simultaneously forces you to stop the grasping and come in close and to connect with where life *is*. The simplicity, the inevitability, the flow, the truth of life. In other words, that Something Else I've gone on about since the outset.

When you've been running scared for a long time this idea may come as a relief. You mean that's *all* I have to do?

Yep, just sit in the grim.

120. Before I go further, I'll bring in our mate Kierkegaard. He used to say much the same, explaining that we spend a lot of our energy running from anxiety. But when we can learn to stay with anxiety, "then the assaults of anxiety, even though they be terrifying, will not be such that he flees from them. For him, anxiety becomes a serving spirit that against its will lead him where he wishes to go."

Poet Rainer Maria Rilke extolled the soul-expanding power of difficulty and urged us to "arrange our life according to that principle which counsels us that we must always hold to the difficult."

What we resist persists.

What we sit in eventually fades to a manageable and livable volume.

When we go low, we come in close and it leads us to the truth of it all.

That's what I reckon.

Kierkegaard also sees anxiety as the very human condition that moves us forward from being mere animals. Worrying about the future has seen us form contingencies and improve our place on the planet. "If man were a beast or an angel, he would not be able to be in anxiety . . . the greater the anxiety, the greater the man," he famously wrote. Charles Darwin, who suffered crippling panic attacks, similarly—and conveniently—claimed that to fret about the future (that is, to be anxious) is to be highly evolved. Kierkegaard, however, adds this clincher to his bold claim, which makes my fluttery heart settle a little as I read it: "He therefore who has learned rightly to be in anxiety has learned the most important thing."

The most important thing? Well, yes. It's to connect with what our anxiety is trying to tell us, it's to go through anxiety to the joy of what just is.

Says David Brooks, not at all in contrast, "The most important thing is whether you are willing to engage in moral struggle against yourself."

Ask yourself, are you? Are you cool to stop running from it, and have the better journey?

121. Some time in the 15th century Zen priest Murata Shukō (Jukō) of Nara turned the tea ceremony on its head. He ditched the fancy porcelain and jade and got "wabi-sabi" with it. Wabi-sabi has no direct translation in English. But the gist is the finding of beauty in imperfection and impermanence, as well as the cycles of messy growth and crumbling decay. This is because that's the way life just goes. And nonresistance IS beautiful. Although the beauty absolutely comes from our nonresistant reaction.

Tea was now poured from clumsy clay and wooden cups with chips of glazing that changed color over time as hot water was repeatedly poured into them.

Through wabi-sabi we learn to embrace our uneven eyebrows, wobbly tables and a nervous need to tap the bathroom door sixteen times (four sets of four) after shutting it.

Because it is what it is.

GET WABI-SABI WITH IT

We can practice finding beauty in imperfection. We can be a bit whimsical and playful with the messiness of life so that we can get closer and closer to it. Whimsy drags us from our purpose-mad existence, it presses "pause" long enough for us to get a taste of life lived in "the now" and freefall for a bit. To see what happens.

Innovation consultant Chris Baréz-Brown writes in *How to Have Kick-Ass Ideas* that ruts are best broken with small moments in whimsy, not seismic changes in behavior. Which is mental muscle building writ differently. Counting men with mustaches on the way to the bus stop is enough to shift perspective, he says.

Candice commented on my blog when I wrote about the topic: "Leave the kids' fingerprints on the wall. I'm choosing to see them instead as a tale of my son's climb up to bed each night."

On forums you find tips like this one: Pick some weeds and play with them until you find a nicely discordant arrangement. Stick them in a jam jar.

Similarly, you might like to cook "fridge surprise" for dinner. Place random ingredients from the fridge that need to be eaten on a plate. Eat them in different combinations, to see how the flavors mix. Anchovies and chard. A mound of left-over pumpkin mash with roast cashews.

Don't clean up the kids' toys before sitting down to dinner with your partner. Have a floor picnic in the middle of it all.

And then just see what happens.

122. This is rather related. It can be a good thing, too, to learn to sit in your own weirdness.

It's also an extension of Kay Redfield Jamison's "individual moments," which we covered on page 170.

When I lived up at the beaches north of Sydney there was this old guy, Bill, who came down to the beach every morning with a butter knife and a plastic bag. He'd sit cross-legged on the grass between the beach and the carpark where the well-heeled locals left their black Range Rovers and Ferraris. Bill was not of the well-heeled set. Nor was he someone you'd describe as intellectually compromised. At all. I stopped one day and asked what he was doing. Sitting in the glorious morning sun, he looked up with a big gentle smile and explained he was methodically extracting a particular weed not local to the area, root by root. It seemed a thankless and endless task. Why did he do it, I asked. "It makes me happy," he said, like it should be obvious.

Bill remains an inspiration for me. I refer to him often. He pays no heed to what "other people" find meaningful or joy-creating. He's worked out what takes him to that place. It's whimsical. It's free.

I generally find that anxious people spend a lot of their lives trying to have fun doing stuff that other people find enjoyable. Things like hens' days, doing big group brunches on Sundays with way too much Hollandaise sauce involved, lying by swimming pools, yum cha, the races . . . actually this is a list of the things that I struggle with. Your list is no doubt different. The point is to recognize that we do this—defer to others' notions of fun. And that this is probably because we struggle with choice (how do you decide what your preference is amid all the things to do in the world?). And to then try to play around with finding stuff that floats your boat. And, no doubt, to then realize that your stuff could be a little weird or unique.

I realize this is a bit weird, but I started working on this—finding out what I liked doing—by signing up to RSVP.com about five years ago for the express purpose of going through the process of filling out the questionnaires that ask you what you like to read, how you like spending weekends and what kind of person you'd like to love you. Knowing you're about to be judged by thousands of strangers gets you really quite focused on getting to the truth.

This is what I came up with: I like talking in tents, catching ferries for an afternoon, sitting at a bar and doing the cryptic crossword, reading 1950s crime fiction and hiking on my own—none of which, I can see now, would be likely to get me a date.

But it did get me focused on acknowledging that I simply don't like doing a lot of what other people like doing. And over time, I got more and more okay with, and less and less anxious about, this.

(PS: Another thing I like doing: noticing and studying weird and unique stuff fellow anxious folk like doing.)

MEDITATE IN GRIMNESS

Off the back of my India visit I started experimenting with meditating in grimmer and grimmer settings. I meditated on planes with kids screaming next to me during take-off. In my office with the team just outside holding a loud meeting. In my car in a parking garage in stifling heat. In a gutter in the sun in central London because I had fifteen minutes to kill and that's where I found myself. In the walk-in wardrobe in the Channel 7 studios, waiting for my turn in hair and makeup before going on to do morning TV.

What happened? I stayed. I didn't try to attend to the ambient conditions—get up to turn down the lights, adjust my underpants and the like—so that I could have the perfect meditation. My only mission was to see what happened if I sat for the full twenty minutes I assign to my daily practice and didn't move. At first the noises, the heat, the smell and even the anticipation of someone walking into the wardrobe and seeing me in such a ridiculous situation distracted me. But each time my mind wandered I simply came back to my mantra, effortlessly, gently. And, you know what, it was almost like the worse the discomfort or distraction, the stronger my focus got. It had to. And what did that do? Well, it saw me go further down into the meditation.

DON'T CHANGE HOTEL ROOMS

It's a thing among anxious types—a need to change hotel rooms. French poet Charles Baudelaire, who was notorious for not being able to sit for long, noted somewhat similarly, "Life is a hospital in which every patient is obsessed with changing beds."

We're uptight before we get to the room. Heck, before we even book the flights and hotel. We know there will be something not right. For me, it will be a hum I can hear from the elevator shaft. Or it will be overlooking a shopping mall's worth of air-conditioning units piled on top of a sprawling rooftop, a cacophony of vibrations that will only worsen throughout the night. I used to change rooms. At least once. Until I worked out that whenever I did, the subsequent room would have worse hums and vibrations, comically so. Wherever I go, there I am . . .

My publisher includes a note at this point in her editing mark-ups—"I experience it often. Once I made [my husband] drive around Port Fairy for two hours with three small children and a dog in the car, trying to find the 'best' place to stay." A few days after she wrote this we're in Melbourne together for a conference. "Did you do it?" I ask her. She laughs, knowing instantly what I'm referring to. "Yep. And the room I'm in now is overlooking the construction site with the jackhammers."

What helps me? I tell myself to try one night in the first room, as an experiment, to see what happens. Again, the meta-purpose of the "experiment" gives me focus. So, too, does the fact that I have an out-clause (I can always swap tomorrow night). When I wake up the next day having slept, I have the courage to do another night in the same room.

SLEEP WITH YOUR PARTNER

Alright, this is a challenge very particular to me. But feel free to confront your own unique one and apply a similar tack. As I mentioned, I avoided sleeping in the same bed with the Life Natural for many months. One night he rendered me choiceless and gave me his most disappointed stare when I was about to run off once again as bedtime approached. I drove away in what I thought was a justified huff. But I got a grip on myself a few hundred yards down the road, turned around and committed to the grimness and seeing what happened. I climbed into his bed and sank into the situation. His Brazilian housemates were having a late-night barbecue outside. He'd not cleared the half-empty beer cans from the bedside table (did I not flag the Life Natural was next-level Australian-surf-culture-laid-back?). I was anxious. I didn't resist it.

He twitched. And snuffled for a bit. I tossed and turned. I watched it all in the not-perfectly-blacked-out room. I didn't expect to look refreshed in the morning. I warned him I might be grumpy.

I got five hours' sleep, as it turned out. And then used the confidence gleaned from having completed the experiment to give it another crack the following week.

ACTIVELY PRACTICE MISSING OUT

My little sister once told me (over Facebook; her preferred family communication medium) that I had the worst case of FOMO she'd seen. Hashtag touché, Jane. It prompted me to experiment a bit. I started to actively plan nights in on my own. I went straight to the grimness of a Friday night with a bowl of fridge surprise (leftovers mashed together in a pan with an egg or some cheese) and *Doc Martin* on the ABC. It sent a massive "up yours" to the ceaseless pressure to be doing and taking part. This emboldened me. A week later I worked all the way through a public holiday. I grimmed it up a treat—I worked from a café that featured Coke-sponsored plastic tables and was frequented only by taxi drivers while everyone else attended Instagrammable barbecues in the sun. It brought on that special feeling of loftiness—not ugly pride, it's more expansive and selfless—that comes from facing and diffusing yet another anxious pocket. High-five me!

grace

123. I think it was a Thursday.

And about seven or eight years ago, during that nebulous, drawn-out period in my mid-thirties. When I was melting down.

In the preceding twelve months I'd left my job in magazines, broken up with the destructive ex, been through the porn star saga, the Machu Picchu and diving-with-sharks misadventures and been diagnosed with Hashimoto's. I'd just been robbed and was contending with the communication breakdowns (phone, internet). I'd been housebound, on and off, between the misadventures, for almost nine months, my Hashimoto's having rendered me unable to walk or work.

I was fat, sick, broke, unemployed, humiliated, isolated, alone, defeated and entirely stripped bare.

Anyway, on this late winter morning I'm kneeling in front of the mirrored wardrobe in my bedroom, silently howling.

I'm clinging to the floor like I'm going to fall through it and I've been scraping at my stomach. It's red and bleeding from the nail tearing. I've been here for three days hyperthinking and knotting my ball of wool, and I haven't slept apart from a dopey hour or two before sunrise each day. The whole sorry story had finally come to a glorious head. I'm in a Category 5 anxious spiral.

I recall thinking that my behavior was all so drama-ish. Did I learn this affectation (the silent howling and stomach scraping) in a movie? I'm a middle-class girl in orange Cookie Monster pajamas having a breakdown on a Thursday when most people are cajoling their kids to eat breakfast or racing to bus stops.

My neighbors are starting their day. The security door slams next to my bedroom window and the kids from the top floor bicker as they clamber down the stairs on their way to school. Life is moving on without me. I've fallen off the conveyor belt. More hyperthoughts.

Perhaps you've never silent howled before? Silent howling is a desperate, primitive scream out to The Gods reserved for when you've sunk as low as you can go. It's that scream from the dank, primordial place I described previously, the pure, chundering-forth expression of the pain at the core of the human experience. It says, *Why?* And *No!* And *Wrong!* It's the outraged scream of a just-born baby after it's ejected from the womb.

It's a silent scream because when you're a highly controlled, insanely well-behaved A-type with a too-tight grip on life, it's beyond you to howl your pain at full throttle. I'm in the worst place I've been in my life. I no longer care about my own welfare. *And yet I'm worried what the neighbors would think if I howled out loud.* And the very fact that I'm micromanaging my own breakdown takes me down even further.

But then, on that Thursday morning, I look up. I look for myself in the mirror. I don't recognize the reflection. I'm gone.

The silent howling stops.

I can only say, looking back, that something dissolved that afternoon. All the things that propped me up and defined "me" had disappeared—my job, my athletic physique, my robust, healthy appearance, my energy, my "strong assertive female" persona, my ability to conceive, my life savings. Gone.

It was a little like that sensation I remember as a kid when dad drove over a dip in the road and my stomach was left in the air. But I didn't bounce back to solid ground again; I stayed suspended.

And this is what I thought in that moment: I'm ready to die.

This was not a violent or desperate thought. Nor was I thinking "I'm currently being suicidal" or "I *want* to die." It just felt like the inevitable dead end of the descent I'd been on. I've since learned that this is what being suicidal can often entail—something of a dissociation from the true meaning of what you are contemplating. It feels like a calm, liberating solution. Dr. David Horgan, from the Australian Suicide Prevention Foundation, says that once we see dying as an option, our minds will focus on finding proof that this is right, ignoring all the evidence that it's a shockingly bad idea.

I'm still looking at my reflection. I have been here a few times before. I know this look. But this time, rather than fleeing from it, I sit with it.

Something wafts into view now. It's an idea and it has words: if nothing matters, if I have no attachments, no commitments and nothing left in my life, I could just quietly disappear. I could self-annihilate. Why not? There was nothing to stop me, nothing I was responsible for. This felt light and liberating.

Or—and now the feeling gets even lighter—I could *choose* to exist, anew. From ground zero, I could opt back in. And I could do it freely, working from a blank slate without all my old stuff—no expectations as to how life "should" be lived, no false and unhealthy ideas about my worth (that I have to achieve to be loved), no attachment to possessions or money. I could be an interloper with no fixed address and just the clothes on my back. I could do life completely differently.

It becomes viscerally apparent that I have nothing to lose and no one to impress. This appeals and it swells as an idea, unhindered by the hyperthinking of yore.

At this point I laugh. It's the laugh I've laughed before when I went bungee jumping in New Zealand with my brother Ben. I fell and I fell, and when I finally experienced the tug at the end of the rope I just laughed and laughed.

And then that was it. I got up, peeled off the Cookie Monster pajamas and stood at the fridge naked and ate peanut butter from the jar with a soup spoon.

124. All of which would be a fairly thuddish anecdote if it weren't for what came next.

The following day I'm at my Eastern medicine doctor, Dr. L, having needles—12 centimeters long—tapped into my ovaries in an effort to inject life, or chi, back into the sad little things. My stomach resembles a mini slalom course and some of the needles are vibrating wildly, doing wide loops over my pelvis. "You are stuck." Dr. L keeps telling me. *I know, I know*.

My phone rings. I answer it. It's TV host Kerri-Anne Kennerley.

I don't know "KAK," as she's known; I'd appeared on her show as a guest a few times.

Lying on the collapsible table and trying not to sound horizontal I realize she's asking me to fill in as host for her show. On Tuesday. "I'm going on leave, something's come up. You'll be wonderful." I don't tell her I've never hosted a TV show before. Or that I don't know how to read an autocue. Or that the day before I was about to take leave on life.

I put on my best Sitting Upright at a Desk voice. "Wow, I'd love to." Because, that's mostly my default position. Yes. Yes, of course. Of this, I'm very grateful. I have been often.

On Tuesday, I played KAK for a day. The Channel 9 wardrobe department dressed me up in a hot pink Fendi dress and bright orange Gucci heels that render me a good foot taller than the infomercial guy who I crack go-nowhere gags with in the second segment. The hair and makeup girls tease out my limp hair and pencil in the gaps in my eyebrows.

Three minutes before going to air I'm shown how the autocue works and told to look to the camera with the red light on top. But I'm not questioning anything. I don't expect anything. A part of me wonders if this is resignation. I'm certainly tired. But I'm also . . . what can I call it?

Loose.

I recall being made to jump on a trampoline that had been pulled into the studio, an advertorial for God-knows-what. And some Taiwanese food sculpture artists created a chocolate carving of my face and bust. They present it to me in the final segment. I'm topless and they'd fashioned me some engorged chocolate nipples.

A few weeks go by and I get a call from Henrie. Henrie was the station's talent director and she'd chatted to me briefly after the show to say I'd done rather well. Shortly after, however, she left her job at the station and moved on to become a casting director. Her first gig was with a new show that was to be called *MasterChef.*

"You used to be a restaurant critic, right?" I had. "I want you to meet the producers of this new show. They might not like you." Henrie has a unique bluntness, one that keeps you from being disappointed by elevated expectations; she was perfect for this period in my life. "But who knows."

I only just remember the audition; my head was incredibly foggy. I recall I didn't know I was being auditioned. I had to taste and critique food sitting on a stool with a bunch of chefs.

I struggled to just get out of the house, my limbs swollen and wobbly from the autoimmune disease inflammation. And I had to borrow a friend's dress because nothing in my wardrobe fitted. I was now a good two sizes bigger than when I left *Cosmopolitan*.

But let's cut now to Christmas Eve. I'm playing Boggle and eating licorice bullets with one of my brothers. Henrie rings. "Darling, you got it." I run outside onto the balcony and I punch the air. *No. Frigging. Way.* "And you're host. You start in two weeks." I punch wildly. I have no words.

125. This is grace. And this, my patient friends who are still with me 286 pages in, is where this anxious journey delivers us.

I'd rather leave aside the standard Christian notion that appears in the Gospels that regards grace as "the love and mercy given to us by God because God desires us to have it, not because of anything we have done to earn it." Although if you reread the above and replace "God" with "the flow of life" then I guess it is in fact pretty much what I'm talking about.

Grace goes a little something like this, and apologies to the philosophers and theologians who've put far more detailed effort over the years into explaining it.

You descend. An anxiety spiral takes you here pretty effectively. In fact, an anxiety spiral is the descent toward grace. Can you see how I'm repositioning things here? Can you see what I've been working up to?

You go into pain. How? You sit in it. You stay. You simply *be* uncomfortable. You get raw. You don't change hotel rooms.

Then you open. As you sit in the pain, you face what you've been fleeing. You see it and let it be. You create space for it to do that old "it is what it is" thing. You let it unfurl and express. This is not easy. But it's bold and brave and purposeful.

Next, you release your grip. You have to. You have to give in. You can see now that you are not the captain of your life. Goddamn, it's hard. But it's the inevitable truth. And with time you do experience it as sweet relief.

Then something shifts. It might be a simple coincidence that presents itself, but one that is so truly random that you have to take note. You know, a sliding door moment. It might be a stroke of luck that turns around your fortunes in one afternoon lying on an acupuncture table. It might be that the openness and humility you've created in this process of going down into your anxiety allows someone to step forward and give you the love you need. They hold you. This, too, is grace.

In all such instances, the sliding door moment or the stroke of luck is not the point. Mostly they're ludicrously innocuous (I mean, a gig jumping on a trampoline in a Fendi dress?!). In fact, I think the stark innocuousness probably helps steer us to the real point—we see that we don't have to do this thing on our own. Yep. That's it. That's grace. Grace is the "is-ness" of life, presented to you on a cracker, ready to eat. It's an openness that plants you into the flow of the river. Grace doesn't bring a party to town. It's not happiness. It's not a fleeting high. It's a delicate, yet whole, gift that whispers in our ear, "Life has this one covered." It tells us that things fit. That you fit. You can't try to get it, you can't earn it or deserve it. It just is. Just as a flower doesn't try to bloom. It just does.

126. David Brooks feels deeply that the endpoint of the anxious journey is the acquiring of character. Writing about the world's greatest thinkers and leaders who pass through suffering before arriving at their significant position in history in the *New York Times*, he suggests, "Many people don't come out healed; they come out different."

I rather love this line. It suggests a subtle transformation or perspective shift, but one that's perfectly pitched for showing you the truth of life. For me I didn't come out healed, I emerged from that touch-and-go Thursday with a calm knowing. A connection. A full, deep sense of the Something Else. A weather vane at my core for what mattered. I also emerged knowing this was enough. It was perfect.

127. In psychological circles it's called post-traumatic growth. I love research that produces a percentage that I can roll out at a party. Like this one: according to the results of more than 300 studies over the past twenty years or so, up to 70 percent of people who went through the anxious ringer report positive psychological growth at the other end. We're talking a greater appreciation for life, a richer spiritual life and a connection to something greater than oneself, and a sense of personal strength. You could call it character.

The way it's described in the literature I've been reading is that certain trauma can shatter our worldviews, beliefs, and identities completely. All the stuff we busied ourselves with—rigidly sticking to a grain-free breakfast for a month, worrying about what our workmates think of our new haircut, resenting that our parents don't respect what we do for a living, getting annoyed that our boyfriends have cooked us dinner once in two months—is obliterated and we're left to start afresh from our real values. The more we are shaken, the more our former selves and assumptions are blown apart and the fresher the growth.

Harvard researchers found this kind of seismic implosion often leads to creativity. The space created by stepping into the "is-ness" of life invites innovative thought and exploration. The examples of this kind of life disaster–first trigger for creative greatness are well known. The research goes as far as showing that people who felt more isolated after a traumatic event reported even greater creativity.

128. After that Thursday morning in front of the mirrored wardrobe, grace paid a few more visits. Not due to any trying on my part. Quite the opposite. It's because I'd given up. I'd gone as far as I could. My life conditions (my grasping, my control freaking, my indecision, my hyperthinking and hypertoggling) had become so intolerable that sitting in the pain, getting raw and open were my only options and so grace just had to flow in. Just as that proverbial "river of life" does when you stop building detours to try and steer it your own way.

I don't think I mentioned that when the porn star got cold feet that time in New York I was already three months into researching and writing the book and had spent my small publisher's advance holing up in Manhattan and L.A. When I got home to Australia the publisher told me I had to repay them the $10,000 advance. It was $9,025.00, to be precise, after a few fees were subtracted. I put up a fight, pointing out it was their talent who'd pulled out of the deal, not me. But I was too exhausted to fire all cannons.

The same week I got a letter from the tax department. I owed a ridiculous sum on some undeclared interest. Well, it turns out the interest was on earnings from a credit union account I hadn't accessed since I was eighteen (I'd assumed it was shut down).

Long. Tedious. Story. Short. I finally get a hold of the obscure credit union in Canberra, from a pay phone at the end of the street (bearing in mind this occurred smack-bang in the middle of that fortnight when I had no phone, internet or car). I'm surprised they're still in operation (both the credit union and the pay phone). They confirm there's an account in my name, but they won't let me access it, nor tell me how much is in it. Not without an account number. Which I guess is fair enough.

Now, I don't remember numbers as a rule. My brain doesn't hold on to them because they don't have a story or a shape. I remember words by picturing their rough form. Which is why I always get Gary and Greg mixed up. But from some dusty recess at the back of my brain I pluck out seven numbers. They tumble out in order, seven numbers I'd used perhaps two or three times more than half my life ago.

They sing-song out like a skipping rhyme. Lachlan the erstwhile humor-free credit union "member services official" whoops down the phone line. "Bingo!" He then pauses like he's a game show host with an unopened envelope. "Oh, yes, there's quite a lot of money in here." How much? I need to know how much. I'm broke. With a tax bill. And a porn star book advance debt.

Lachlan can't tell me.

I have to supply a signature. Again, fair enough. We tried my signature, which I faxed to them from the post office down the road. It took three failed attempts to actually physically get there (remember, my car had just been stolen). But the signature didn't match. I vaguely recall redesigning my signature in my late teens.

Carless, cashless and terribly weak I go back and forth between various agencies until Lachlan and I are able to finally reunite (again, via the pay phone) and he can *finally* tell me how much bloody cash I have sitting in this mystery account, that's been ticking over year after year, growing exponentially and accruing tax debt at the same rate.

"You ready for it?" Lachlan pauses dramatically again. "Miss Wilson, it's $9,025."

This was a very fine moment in grace.

129. There's a very particular thing about grace. You can't go out and get it, or buy it. Just as you can't earn it. I read somewhere that it's like life wants to give us a gift, but we want to buy it.

Remember Sky? The spiritual counselor charged with making sure I didn't get too caught up in the magazine world I found myself in? Well, in our final session I called on her to help me quit my job. I fretted back and forth with her over many sessions. Should I dump everything and enter the unknown? Or should I wait for a better job first? Or should I just persevere? Because what if I had this wrong? What if life really was about getting a secure footing on the conveyor belt and neatly passing from school to job to partner to holidays in

Fiji for two weeks every year to bridge nights? And so on and so forth.

I've asked this question so many times in my life. I asked it when both my previous long-term relationships came to an end. I fretted whether I'd ever find anyone better. I went through it when I deferred university to travel for a year. What if I was wasting a year in which I could be getting ahead? Our default position is safety. A desire to buck it gets messy.

Sky let me fret and squirm for a bit before sharing this profound wisdom that I've referred to many times since.

"The thing about life, sweetheart, is this, when we leap into the unknown, we always land safely. We just do.

"We freefall for a bit." She does a zooming thing with her hands. "But then, as we're falling, we grow angel wings that carry us to our destination."

I can't quite believe she's introduced angels without apology, but I nod.

"Life supports us; it always does.

"The problem is, we all want to go out and buy ourselves a set of angel wings *first*. Before we jump." She nods at me to check I'm getting her drift. I am.

"But, sweetheart, there's no such thing as an angel wing shop."

There most certainly isn't. You have to jump first.

And, you see, that's the other thing about grace. You have to let go first.

In our culture, we want guarantees. When we can learn to make leaps without them, then, well, things really do start to look different.

I go home. I put Gillian Welch on my stereo and type out my resignation letter. It's a cracking email. There's no room for equivocation, no suspended clauses that allow for my own backpedaling. I write, "It is with clear and irrevocable intent . . ."

GET OLD

I chatted with Mitch Albom, he of the mega-selling *Tuesdays with Morrie*, during one of my New York trips. He shared with me this idea. "When a baby comes into the world, its hands are clenched, because a baby, not knowing any better, wants to grab everything, to say, 'The whole world is mine.' But when an old person dies [it's] with his hands open. Why? Because he has learned the lesson."

I wonder if sheer years on the planet is the ultimate balm for anxiety. In those letters to one's younger self that magazines and chat forums like to do with famous or well-lived people, the advice handed down is always that "it gets better. It just does."

I raised this in the forums I held while writing the book. Participants aged in their teens and twenties arced up about the idea. They didn't like being told that working on their anxiety would help and bristled around the idea of sitting in anxiety, accepting it, seeing the beauty in it. In one session we talked about a radio program some of us had heard on the way to the session. It discussed resilience. A listener shared her extremely traumatic childhood and was asked what stopped her from living with the "it's unfair that I got lumped with this" noose around her neck. Her simple answer—hindsight. She was now thirty-five and, she said, she'd simply had enough rough patches from which she emerged okay. These eventually stack up and create a picture. That life will turn out okay. One 24-year-old participant in the forum blurted, "It makes me really angry . . . it's bullshit and it makes me more anxious. So does your talk of 'gratefulness.'"

The participants over thirty-five or so had a different response.

> *I've arrived at an age where accepting this is "just my life" brings peace and, going through the motions of anxiety when it arises, strangely it helps. This too will pass. You fight it still, but it lessens over time.* – Anthea

> *I followed the "right path," doing all the "right things" to keep anxiety at bay. But it didn't work. After 20 years you let go. Having toddlers are good [sic]–they do the opposite. You have to let go and give in or you will be one of those people whose bodies collapse.* – Tim

As Steve Jobs shared in his Stanford commencement speech, life and its hardships only make sense when you get old enough and you're able to look back and join the dots. You have to have dots in your experience for the picture to take form. When you look back on your life you can see that pulling out of college from nervousness (as he did, instead sitting in on typography lectures), is what led you to run a graphically orientated tech empire. But only once you have enough dots.

130. Jump first. A bit of an ask, hey. Does life realize what it's asking of us? I think it does.

If we're serious about joining life–like *really* joining it and not sitting at odds with its flow and existing constantly in a state of dis-ease—we gotta have faith. (I'm aware some of what I'm saying here is starting to sound religious. Please don't leave me if this presses buttons for you. Before their scriptures turned into dangerous, numb doctrine, most of the spiritualists asked the same true questions and used the same language. That's all.)

Life is mysterious. Life is uncertain. We don't know what's going to happen. Along with taxes and death, the only certainty in life is that we just don't know. So we might as well join this inevitability. In my chat with Brené Brown we discussed that this is the ultimate way to live a wholehearted life—to get cool with uncertainty (for the record, we also discussed how very little frightens us more). Sartre described it as a necessary experience that allows us to "become free in relation to our nothingness."

Many psychs today discuss managing anxiety in terms of having "negative capability." Which is to say, having an ability to be okay with the uncertainty of life. The term emerged from a disagreement John Keats had on the way home from the Christmas pantomime with a bunch of contemporaries who were on a professional quest for definitive, reductionist answers. Irked by the notion, Keats writes:

> *Several things dovetailed in my mind, & at once it struck me, what quality went to form a Man of Achievement especially in Literature & which Shakespeare possessed so enormously—I mean Negative Capability, that is when man is capable of being in uncertainties, Mysteries, doubts, without any irritable reaching after fact & reason.*

What an aim. To sit comfortably in mystery without grasping outward. To sit. To stay. And see what happens. It's freedom, right? But how do we do it? Dare I say it, I think it takes patience and sheer years on the planet. Rilke writes in *Letters to a Young Poet:*

> *I beg you, to have patience with everything unresolved in your heart and to try to love the questions themselves as if they were locked rooms or books written in a very foreign language. Don't search for the answers, which could not be given to you now, because you would not be able to live them. And the point is to live everything. Live the questions now. Perhaps then, someday far in the future, you will gradually, without even noticing it, live your way into the answer.*

GO STRAIGHT TO COOL

It can be a scary chore to set out and "trust life." So I take a slightly different tack.

I go straight to being the person who is open and cool with not knowing. I practiced this heavily while coping with the vagaries of my disease and my diagnosed infertility and (somewhat resulting) singledom that lasted eight years. I kept saying to myself and others who asked what the future held, "I don't know." But I wouldn't say it despondently; I'd be deliberate about being cool with it. In doing so, I found a strength that is quite defining and satisfying.

It meant my vulnerability was about being raw and exposed, but ultimately was something I steered and owned. Over many, many years of building this muscle, I now feel emboldened when I say, "I don't know."

131. American Buddhist nun Pema Chödrön (who cites her two marriage breakdowns in her twenties as the catalysts to her own spiritual and anxious journey) defines anxiety as resisting joining the unknown. I came across one of her books wandering around a bookshop where I wrote a lot of this Beautiful Beast. Like Ray Bradbury, who wrote *Fahrenheit 451* in a library in twenty-minute spurts broken up by flicking through random books that (gracefully?) provided perfect inspiration for his seminal book, I flicked randomly, loosely, finding the sport of seeing what cropped up very helpful. Turns out, as I grappled with this very bit you're reading, I realized the book I'd picked up was called *Comfortable with Uncertainty*. What are the chances? Better still, the subtitle: *108 Teachings on Cultivating Fearlessness and Compassion.*

— Yep, 108 again. It's an auspicious number in several Eastern traditions and is a mathematically pure and abundant number. Plus, the diameter of the sun is 108 times the diameter of the Earth. I like all these factlets. And that the number pops up in my life daily.

Chodron argues that the journey we all need to do is the experiment with sitting in uncertainty. *Ha!* The ultimate endpoint, she writes, is growing up. The journey "offers no promise of happy endings." Rather, the part of ourselves that keeps seeking security (when there isn't any) and something to hold on to (when such a thing doesn't exist) finally grows up. She says anyone who faces these truths is a true warrior.

the last bit

132. I've always loved the parable of the Little Speck of light that lives in the middle of the sun. I've long forgotten where it comes from, or who first shared it with me. The Little Speck calls out to God that she's ready to find out who she is. So God takes the Little Speck and deposits her far out into the darkness of the universe. There the Little Speck is surrounded by pitch black, which freaks her out. Against the darkness, not surrounded by the other familiar specks of light in the sun, she sees herself alone. She cries out to God, "Why have you forsaken me? I wanted to see who I am! This is not what I asked for."

God says, "To see that you are light you must first go out into the dark."

To see yourself—to see that you are part of a big, magnificent whole—you have to go to the depths. I believe this now more than ever.

133. We get anxious if we feel we're not connected with our true selves and what matters. Something is not right, something is missing, we don't understand what life is all about, and this gnaws at us. It doesn't have to get particularly woo-woo. It plays out as a general social uneasiness (which we think we'll just get the knack of eventually), as a sense that we haven't got to where we're meant to be (in our career, in our love life), as a persistent cynicism that it's all a facade (the white picket fence imperative, the smug dinner party talk, the sea of selfie sticks), as inflammation in our bodies, as a desperate need for more food, more wine, more friends, more likes, more throw cushions. As dysfunction.

We're unsettled, we grasp and we grasp. It's like we're bobbing for apples in a big barrel of water. But we come up with nothing.

But—oh glory be—by being in anxiety, by going down to the dark depths, we finally find the connection. Because anxiety, eventually and inevitably, makes us sit in our shit. It takes us there, to the darkness. It forces us to do the journey. And only then can we see what we were looking for. We can see the truth. We see it all as it is.

Anxiety makes us come in closer and eventually we arrive at something particularly and unspectacularly un-woo-woo. Ourselves. Phew, hey. *Phew*. The exhausting outward chase can stop. It's all here. Right here. No need to run, anymore. We let go; we join the flow of life. It makes sense; we belong. Because that's all there is. That's what anxiety does for us. It guides us home.

And when we veer or we deviate from the truth, anxiety steps in and forcibly tells us "Wrong Way Go Back."

Can we find untold wisdom and maturity and meaning without going through the ringer of anxiety?

I guess so. But would we go there unprodded, knowing the grit it takes? I doubt it. As W. B. Yeats wrote, "It takes more courage to examine the dark corners of your own soul than it does for a soldier to fight on a battlefield."

I've been following the ironies and confusing paradoxes that anxiety presents. They've always seemed twistedly cruel and unfair. And my head has spun to understand, why me?

Why has life forsaken *me*?

I dunno, I guess it's because I did cry out (to the Gods?) some time back to *know*. I wanted to know what the hell this is all about.

Be careful what you ask for, could be the lesson here. But it's not.

The lesson is that my anxiety is what delivered me the answer. Which is the ultimate irony.

Eighty-odd years ago, Freud proposed that anxiety was "a riddle whose solution would be bound to throw a flood of light on our whole mental existence." The process of trying to solve the riddle, he reckoned, would help us unravel the juiciest mysteries of the mind: consciousness, creativity, pain, suffering and hope. To grapple with anxiety is to finally understand the human condition.

Have I solved the riddle?

Goodness, no. But I get that it's all a riddle. It's a wrangling with the truth that gets us closer, stronger, more warrior-like. We can view anxiety as something to accept and live with. Sure, this is important. But I reckon we can make the beast more beautiful than that. I prefer to say (to quote Shai from one of the forums again) "anxiety is my superpower."

134. Along this bumpy path, I've come to think that my problem was not so much that I had a problem. No, my problem was that there was so little guidance available to steer me as I wrangled with my anxieties, particularly in those formative years in my late teens and early twenties when many unhealthy mental patterns develop. There were no sturdy arms to hold my buzzing energy. No social framework that assured me there was a place to come back to when I'd descended. No wisdoms. No sitcoms that showed people sitting in their dark, fretful anxiety and getting really truthful with themselves. No one is to blame for this. It's just that things have not been discussed in this manner for a very, very long time.

This, though, is the conversation we now need to have. This is the level of discerning, mindful, connected dialog I think many of us are crying out for. We need to discuss the riddle.

Kierkegaard reckoned that anxiety is an "adventure that every human being must go through." And Socrates said, "An unexamined life is not worth living." I disagree. Not everyone wants the adventure. Not everyone *wants* to lift the scab. But those of us who do need the new conversation.

135. Writing this book I was kept in real-time anxiety. It took almost two years to write. Indecision and a liberal peppering of anxiety spirals delayed the process. Heady injections of mania ricocheted me back again to the page (although entire slabs of manic outbursting had to be removed on re-reading). During this time, I flitted between nine countries, moved house seven times, attempted suicide twice, restructured my business and fell in love with a man who pushed every single one of my fretty buttons. He held one of those shaving mirrors with an embedded fluorescent light that they put in hotel bathrooms up to my filthy-mitted grip on life. The reflection was oversized and all the scars were visible.

I sat in this challenging reflection and loved him more for taking me to the grim depths. I really did. We spent weekends in the garage under the group house he shared with a revolving door of South American backpackers "hanging out" (I'd never "hung out" in my life; the notion filled me with a lawnmower-drone-on-a-Sunday-morning dread). Me reading the newspaper, researching, sometimes knitting; him painting and fiddling and watching surf reports on the iPad he'd set up for such purposes and listening to the football on the radio, drinking cans of beer. When we felt like it, we got up and jumped in the ocean. And pondered where sand came from. Many around us couldn't work out how we fitted together.

Other times every bit of my deep pain railed against him, tested him, and I twisted the dialog into a double-helix spiral. I was railing against my own fear of letting go, of course, testing myself more than him. His reaction was to flee.

Then we got pregnant.

After eight years grappling with being a barren woman—watching friends produce beautiful children with increased frequency; telling men on the third date that I couldn't have kids (then watching them fade away after a week or two); coming to terms with the loss of never having access to that maternal softness (with both a baby and my own mother)—this was the kind of rare shock that shut me up.

There were no words. I couldn't tell close friends. I rationalized everything in my head, but inwardly broke down from the bigness of it all. To the Life Natural and others who did know, I appeared stoic, I guess. I'd planned my life around protecting myself from the barren-woman status. I'd built a career around helping people so that I could have something to show for myself when I arrived at my deathbed with no grandchildren. And now, and now . . . frankly, it was all out of control.

The Life Natural had intuited the pregnancy in his simple, connected way. "I think I just got you pregnant," he said the night we conceived, knowing the biological impossibility of it (it turns out it could only have been that night; I'd had an ultrasound the day before and got on a plane to London the next morning for two weeks). It all made sense to him. He loved the magic that had just happened.

I came to see the magic—the grace—of it, too. And a funny thing happened. About six weeks into the pregnancy I realized I wasn't anxious. There was no background buzzing. It was wonderfully odd and I reflected that this was the only time I could recall not feeling anxious in more than twenty years. It was the oxytocin; it was the progesterone; it was the flooding of resources to new life in my belly; it was pregnancy brain (for some it makes them forgetful, for me it brought my thoughts into line with everyone else's); it was love.

I knew when I'd miscarried, at ten weeks. The lovely, centered fullness had gone one morning and the brittle, noisy hollowness had returned.

I'm so sad to say the pain ate away at us. We trudged our way through a protracted anxiety-riddled breakup (or breakdown), prodded periodically by big doses of mania as I tried, relentlessly, to save us. The Life Natural had never encountered such complexity. This isn't you, he'd say. The next moment he was telling me I was a bitch. My behavior was certainly bitchy. He couldn't reconcile I was neither. Or both. Or more than it. He couldn't understand that I wasn't trying to control him; I was trying to control my terror. That my attacks on him (which were wild and desperate) were pleas for help.

It took me months to realize that he was wholly bewildered. I tried to make it simple. I was a big wave, I told the Life Natural. The biggest he'll ever encounter. He sought out big waves, he traveled the world to lunge himself at them (the non-metaphorical kind), testing life much as I like to. To ride a big wave you first have to paddle out hard to get behind the point where they start breaking. You have to fire up and give it your all. If you do it half-heartedly, you'll get smashed, and you'll be pushed back into the churning whitewash over and over.

I told him that if we get out the back, we could ride the big energy together. In to the shore and back out again, leaving the whitewash behind.

"All you have to do is join me. Hold me. I'll take care of the rest."

It was an invite, admittedly developed with manic enthusiasm and probably repeated too often to have any potency after a while. The Life Natural disappeared for a few weeks, then emerged to tell me my anxiety was too big for him. He couldn't ride with me. He was deeply sorry.

This remains the hardest, most unresolved part of the anxious journey for me, and for many I've had the conversation with. The journey has to be done on your own. This is a terribly lonely thing to have to live with. And the loneliness hurts like hell, out there in the dark on your own.

So what was the reflection the Life Natural held up to me in that obnoxiously lit mirror? That if I want to let go, to truly let go and trust life, I first have to let go of the idea that someone else must hold me while I do it. No one else can tell me that life has this one. I have to do this for myself.

And back on the road I go.

136. Some pretty recent anxiety research focuses on how resilience allows anxious folk to "thrive despite their anxiety." One study conducted by Dennis Charney that got quite a lot of media attention (heightened in late 2016 when Charney was shot by one of his former researchers) identified ten factors that create resilience, among them having a moral compass or set of beliefs, faith and spirituality, an ability to leave your comfort zone and face your fear, having a sense of meaning in life and having a practice for overcoming challenges. I read about this research in the final week of writing this book and it made me feel less nervous about sending it to my publisher.

But I'd put it to Charney that it's anxiety that leads us to these factors. Indeed, I'd say anxiety creates the resilience to thrive in this life. Anxiety is a beautiful thing.

137. Did I come out healed? Will I ever?

Since childhood I have cried out to know where I fit, for life to make sense, to learn how to sit comfortably with myself on that bench in the sun. After thirty-odd years of doing the damn journey, have I arrived somewhere? Anywhere? Is that the question you now ask me?

David Brooks wrote that those who embark on the road to character as he puts it, or the path to meaning and sense, as I'm putting it, don't come out healed. They come out different.

I don't sit here healed. I sit here simply knowing I'm on a better journey. And this is enough. This is everything.

When my anxiety gets bad, I stay with the pain. I don't flee; I ride it out. I watch it. I cope. The rest of the time, I prepare, I buttress, I loosen the knots, I modulate, I build muscle with little right moves. And all the time I'm coming in closer, I'm understanding. I'm having the better journey.

I still meditate badly. But every single day that I sit down to do it, little pockets of tension bubble up that make my head jerk violently and I know I'm releasing my anxiety, little by little. And that folding forward thing I do each time at the end of meditation? Where I take my energy into the day? And I hold it and hold it? That does great things for me.

I've let go of some friends and peers; anxiety has alerted me to when this, sadly, has to happen. I've watched, astounded, as relative strangers come forward to help me *exactly* when I need it. I know how to spot good people, around the world. And I write letters to them and book in times on Skype to have discerning, mindful conversations with them.

I accept some people in my life feel better believing that I have a chemical imbalance in my brain. I take pills to sleep sometimes. I stumble, but like in yoga when you fall over in flying eagle pose, it's the stumbling that sees my muscles unify and my mind become still and forgiving. Because it must. That's where this better journey takes you, into a space where you keep being rendered choiceless and forced to soften.

I'm more mature. I like the stillness and certainty that comes with sheer years on the planet. I know that twenty minutes of swimming across an ocean bay will get me grounded. I know that all I have to do is tie on shoes, stuff a credit card and my phone down my bra, and take off to some dirt and rocks and hike when I have anxiety in my bones. These two things—ocean swimming and hiking—are what make me the happiest. And anxiety brought me to them.

I notice coincidences and don't place too much importance on them. I just find them funny, like the fact that a moody early-summer Sydney storm has erupted as I write these final lines. I can say "I love you" and I know that I love loving.

I'm getting better at knowing what to care about. Again, anxiety is my compass. If I'm anxious, I know I'm going the wrong way.

I can deal with trolls and bullies. I see their barbs as a tennis ball flung my way. I can put energy into belting it back. Or I can let it sail past and land flaccidly somewhere behind me. I absolutely don't care and this fascinates me.

I love Feist's "I Feel It All," because I do (feel it all).

I'm lonely and I'm awkward around a lot of people. But I enjoy talking to strangers who do unique, whimsical things. Like the old couple in the park who paint little canvases of whatever they see from whatever bench they sit on. This is the best thing for making you feel less alone.

I am anxious often. But it's kept in check if I don't get anxious about being anxious. And while ever I'm learning more, understanding more, this is entirely possible. Yep, the journey is what matters most. It's everything.

This morning I went for a crazed run around the harbor and back up the steep stairs and now I'm eating 85% dark chocolate. I want to feed the anxiety. To speed things even further. As I do, I'm watching myself Doing Anxiety. It passes after a bit. I make some turmeric milk. I eat some leftover chicken and cheese and avocado and coriander for lunch.

I love that perfect quotes emerge like pink Volkswagen Bugs when you put your mind to a topic. Like this one that I found, again, while wandering around the bookshop in a writing break, stuck on this idea of whether I'm healed or will ever be: "Give up on yourself," wrote Japanese psychologist Shoma Morita. "Begin taking action now while being neurotic or imperfect."

Yep. Just do it.

Now, mostly my joy comes from knowing I can do both. I can be neurotic and imperfect.

The camera is still rolling.

For the raindrop, joy is in entering the river/
Unbearable pain becomes its own cure.

— 18th-century Urdu poet Mirza Ghalib

acknowledgments

I couldn't decide what order to thank people in. So I cut up all the individual thank-yous from the Word document and pulled them out of a saucepan. In no particular order (for I'm grateful equally in all directions):

Thank you Damian for generously, uniquely allowing—and encouraging—the story to be told.

Thank you everyone who let me quote them, share their story and paint pictures with their ideas. Including all the readers on my blog and strangers who talked to me about their anxiety, and those who took part in the SANE and Beyond Blue forums. Thank you for your wisdoms.

Jo, you carried the load, you got me from the hospital, you never judged me, you are loyal to the end and I'm indebted. Rick and Brad, you held me and you housed me through the worst of it and when I threw a tangled tantrum with your clothes horses you only told me you'd done exactly the same and that clothes horses are bastards. Helen and KJ for doing the right thing. Your compasses are true. Ragni, Kerry, Kersti, Faustina, Tim M, Claire, Lizzy, Paul, Bill, Annie, Tim B. and Nicho for listening to me. And accepting it all.

Ingrid, you gave me so much rope and understanding and steered the book without ever forcing it or me. Georgia and Miriam, I'm ever grateful for your involvement. And Miriam, I thrived on your comments in the side column as you edited ("Choose one metaphor darls, otherwise it sounds like a night at the Hellfire Club!").

Ben, Pete, Si, Jane, Nick. Just because my love for you has been what's kept me here at times.

And thank you to the rest of the Pan Macmillan team, and designer Daniel New, not because one must always thank such folk in the acknowledgments section, but because you really did go on the ride with me. You let me have false starts, you let me be.

Thank you to my whip-smart and caring U.S. agent, Stacy Testa, and to the wonderful team at Dey Street Books—Jessica Sindler, Maria Silva, Kendra Newton, Benjamin Steinberg, and Lynn Grady—for giving a U.S. home to this book.

five extra reads you can find on my website*

1. A full list of science and source endnotes from my book
2. A full list of my recipes for meals that help modulate anxiety
3. A full reading list of great anxiety books by mindful types
4. Treatments that have helped my anxiety and where to find them
5. Good anxiety apps and sites you need to know about

* sarahwilson.com . . . or you can simply whack into Google the name of the read and my name.

text acknowledgments

Extract on page 6 taken from *The Fry Chronicles* by Stephen Fry, published by Penguin Books Ltd. Reprinted by permission.

Extracts on pages 6 and 34 taken from www.thebookoflife.org by Alain De Botton. Reprinted by permission. All rights reserved.

Extract on page 12 from "The Journey" from *Dream Work* by Mary Oliver. Copyright © 1986 by Mary Oliver. Used by permission of Grove/Atlantic, Inc.

Extract on page 32 used by permission of Prof. Matthew Walker, UC Berkeley.

Extract on page 75 taken from *The Balance Within* by Esther Sternberg, MD, Research Director, Arizona Center for Integrative Medicine, Director, UA Institute on Place and Wellbeing, University of Arizona College of Medicine, author of *Healing Spaces* and *The Balance Within*. Used by permission.

Extract on page 76 taken from "The Grateful Brain" by Dr. Alex Korb, PhD, accessed via www.psychologytoday.com. Used by permission.

Extract on page 99 from *The Three Marriages* by David Whyte, published by Riverhead Books. Reprinted by permission of Penguin Random House.

Extract on page 103 from *Jung Collected Works 22 Vols*. Reprinted by permission of Routledge, Taylor & Francis Group, UK.

Quotes on pages 112 and 222-3 taken from "Louis C.K. Is America's Undisputed King of Comedy" by Andrew Corsello, published in *GQ*, 13 May 2014. Reprinted by permission.

Extract on page 164 from *A First-Rate Madness* by Nassir Ghaemi, published by Penguin. Reprinted by permission of Penguin Random House.

Excerpts on pages 167 and 201 taken from *The Diary of Anaïs Nin*, Volume Four: 1944–1947. Copyright © 1971 by Anaïs Nin. Reprinted by permission of Houghton Mifflin Harcourt Publishing. All rights reserved.

Extracts on pages 169 and 170 taken from *An Unquiet Mind* by Kay Redfield Jamison. Copyright © 1995 by Kay Redfield Jamison. Reproduced with permission of Pan Macmillan via PLSclear.

Extracts on pages 192 and 274 taken from *The Road to Character* by David Brooks, published by Random House Publishing Group. Reprinted by permission of Penguin Random House.

Extract on page 253 taken from *Mindfulness for Unravelling Anxiety* by Richard Gilpin, published by Ivy Press. Reprinted by permission.

Grateful acknowledgement is also given to Glennon Doyle Melton, Hugh Mackay, Ira Glass, John Koenig, Junia Bretas, Matthias Mehl and Ruth Whippman, for granting permission to be quoted in this book.

For a full list of resources referenced in this book, please visit **sarahwilson.com**.

where to find help

If you're in a bad way, call professional help. It's imperative.

A good place to start if you're in an anxiety spiral or having a panic attack . . .

Crisis Call Center: This 24-hour crisis line is open 7 days a week, 365 days a year, and offers emotional support for those in crises: www.crisiscallcenter.org/crisisservices-html/. Call 1-775-784-8090. Text: 839563

National Suicide Prevention Lifeline: Another 24-hour lifeline that provides free and confidential support for those in emotional distress or contemplating suicide: www.suicidepreventionlifeline.org. Call 1-800-273-8255.

If you suffer from anxiety as a persistent thing, I can't overstate the importance of putting the work into digging around and understanding your fretting better. Know this: on average, it takes an anxious person six years to find a counselor or psychiatrist who floats their boat. Me? Yes, it took about that long to find the help that made a difference. And even then, I found it useful, after a few years, to move on to a different discipline or professional who could challenge and guide me in different ways. It's all part of the process; you learn from the not-so-good shrinks, too.

Anxiety and Depression Association of America: A nonprofit dedicated to helping people better understand anxiety and depression. They have a directory of therapists specializing in anxiety, depression, OCD, PTSD, and related disorders: www.adaa.org/finding-help

PsychCentral: Run by mental health professionals, this site connects readers to online mental health resources and articles. www.psychcentral.com.

National Institute of Mental Health: Offers research information on mental health disorders. They also list clinical research studies: www.nimh.nih.gov/index.shtml.

National Alliance of Mental Illness: A grassroots organization dedicated to improving the lives of those affected by mental illness through educational programs and advocacy: https://www.nami.org/About-NAMI.

Mental Health America: The nation's leading community-based nonprofit committed to promoting mental wellness through prevention, treatment, and education services: www.mentalhealthamerica.net/.

Canadian Mental Health Association (CMHA): Through advocacy, education, research, and service, this volunteer organization provides support to more than 1.2 million Canadians: www.cmha.ca.

AnxietyBC: This online charity offers online, self-help, and evidence-based resources for Canadians affected by anxiety: www.anxietybc.com.

National Network for Mental Health (NNMH): A nonprofit that connects Canadians to information and experts on anxiety and other mental health disorders: www.nnmh.ca.

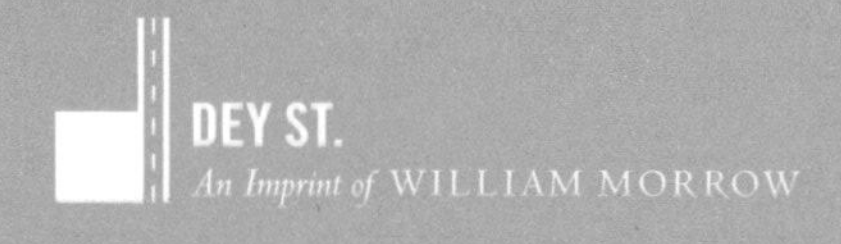

Cover: Daniel New / OetomoNew
Cover illustration: printmyfish.com
Original cover and internal concept: Sarah Wilson